COASTAL PROCESSES

ADVANCED SERIES ON OCEAN ENGINEERING

Series Editor-in-Chief
Philip L- F Liu (*Cornell University*)

Advanced Series on Ocean Engineering — Volume 28

COASTAL PROCESSES

Concepts in Coastal Engineering and Their Applications to Multifarious Environments

Tomoya Shibayama

Yokohama National University, Japan

 **World Scientific**

NEW JERSEY • LONDON • SINGAPORE • BEIJING • SHANGHAI • HONG KONG • TAIPEI • CHENNAI

Published by

World Scientific Publishing Co. Pte. Ltd.

5 Toh Tuck Link, Singapore 596224

USA office: 27 Warren Street, Suite 401-402, Hackensack, NJ 07601

UK office: 57 Shelton Street, Covent Garden, London WC2H 9HE

British Library Cataloguing-in-Publication Data
A catalogue record for this book is available from the British Library.

COASTAL PROCESSES: CONCEPTS IN COASTAL ENGINEERING AND THEIR
APPLICATIONS TO MULTIFARIOUS ENVIRONMENTS
Advanced Series on Ocean Engineering — Vol. 28

Copyright © 2009 by World Scientific Publishing Co. Pte. Ltd.

ISBN-13 978-981-281-395-4
ISBN-10 981-281-395-0

Printed in Singapore.

Preface

This book was originally written as a textbook for a course of coastal engineering given at the Asian Institute of Technology in Bangkok in 1991, where I described important concepts in Coastal Engineering and their applications to coastal processes. A part of the book describes basic concepts of coastal engineering that deal mainly with wave-induced physical problems. The other part consists of my scientific research results for the past these 30 years. For the case of sediment transport study, the book covers not only coastal zone but also sediment production in river basin and river sediment transport in order to understand the mechanism of coastal erosion. From the contents of this book it is also possible to understand the different situation of various countries regarding this problem.

Since I have spent most of my working years in Asia, my research interests are strongly influenced by Asian coastal problems, particularly coastal erosion processes and natural disasters. The book clarifies the macroscopic relationship between the industrialization process of developing countries and coastal erosion or deposition process based on the examples of Japan and other Asian countries. The book also gives microscopic detailed understanding of the dynamic process of sediment production and transport from river basin to coast.

The book also provides a way of understanding and develops effective countermeasures to coastal disasters such as tsunami, storm surge and high wave attack. The descriptions are based on my own observations and measurements of disasters caused by for example, the Indian Ocean Tsunami in 2004, Storm Surge of Hurricane Katrina in 2005 and Central Java Tsunami of Indonesia in 2006.

Selected research results of my former students are introduced in the book. Most of them graduated from Yokohama National University with their doctoral degrees in the field of Coastal Engineering and are now working in their home countries. They are; Dr. Nguyen Ngoc An, Dr. Nguyen The Duy and Dr. Nguyen Danh Thao at HoChiMinh City University of Technology, Dr. Winyu Rattanapitikon at Thammasat University, Prof. Li Shaowu at Tianjin University, Dr. Mohsen Soltanpour at K.N. Toorsi University of Technology, Dr. Ioan Nistor at University of Ottawa, Dr. Nimal Wijayaratna at Ruhuna University, Dr. Kweon Hyuck Min at Kyonju University, Dr. Jayaratne Ravindra at University of Liverpool, Ir. Masimin at Syah Kuala University, Dr. Michael Kabiling at Taylor Eng. Inc., Dr. Wudhipong Kittitanasuan at Wishakorn Consultants, Dr. Le Trung Tuan at Vietnamese Institute of Water Resources, Dr. Le Van Cong at Vietnamese Academy of Science and Technology, Dr. Hiroyuki Katayama and Dr. Manabu Shimaya at Penta Ocean Research Institute, Dr. Hiroshi Takagi at Yokohama National University, Dr. Takayuki Suzuki at Port and Airport Research Institute, Dr. Joel Nobert at University of Dar Es Salaam, Dr. Miguel Esteban at United Nations University, Dr. Hendra Achiari at Bandung Institute of Technology and Dr. Thamnoon Rasmeemasmuang at Burapha University. Dr. Jun Sasaki, my colleague, also supported field surveys of coastal disasters in these five years. In the editing process of the book, I acknowledge the assintance of Ms. Akiko Nakao, Dr. Miguel Esteban, Dr. Nguyen Danh Thao and Ms. Yuko Tachibana.

Finally, I would like to extend my sincere gratitude to Dr. Makoto Shibayama, Professor of Development Psychology at Kamakura Women's University, my wife, for her valuable comments in writing this manuscript.

March 2008
Tomoya Shibayama

Preface of Old Edition

This book is originally designed for the lecture series on "Basic Coastal Engineering" which will be held in Indonesia on August 1991. The participants of the lecture will be university instructors who are teaching coastal engineering courses in undergraduate and graduate levels of Indonesian universities. In preparing this manuscript, I have set the aim to write a more general and basic textbook on coastal engineering. Since I have spent most of my working years in Asia, my research interests are strongly influenced by Asian coastal problems, particularly on coastal erosion and deposition processes.

The contents of this book deal mainly on wave-induced physical problems in coastal zone and coastal processes which are designed to give basic theoretical and practical materials to undergraduate and graduate students whose major field is coastal engineering. Part I is basic coastal engineering for undergraduate and graduate students while Part II is a collection and a reprint of my recent research papers. Part II is designed for graduate students pursuing advanced study in coastal engineering. In order to complete the course on basic coastal engineering (Part I), a student is required to spend thirty hours in class and additional thirty hours in homework.

Portions of Part I are based on the lectures of Prof. K. Horikawa and Prof. A. Watanabe of the University of Tokyo and Prof. O. S. Madsen of Massachusetts Institute of Technology (MIT) who were the author's instructors during his graduate studies. They taught me the fundamental principles of coastal engineering. I would like to thank Professor Fumio Nishino who gave me a chance to work at the Asian Institute of Technology in 1990-1991 and 1983; and for giving me the opportunity to deliver lecture series in Indonesia in August 1991. Concerning Part II, I would like to thank my colleagues, especially Dr. Shinji Sato and Dr. Akio Okayasu who worked with me for the past ten years. I also wish to acknowledge Ms. Isolde Colinares Macatol, AIT, for her assistance in editing this book. Finally, I would like to extend my sincere gratitude to Makoto Shibayama, my wife, for her unfailing support and inspiration that has helped me greatly in studying and writing this manuscript.

Asian Institute of Technology, Bangkok
August 1991
Tomoya Shibayama

Contents

Chapter 1

Introduction

The term *coastal engineering* had been used for port construction design and then extended to disaster prevention in the late 1950's, and to environmental protection in the 1970's. Its purpose is to solve engineering problems in construction works under coastal and ocean environments. Figure 1.1 shows the general view of areas that coastal, ocean engineers and oceanographers deal with. Initially, the area of discipline of coastal engineers was limited to the shallow-water region, however, with the advancement and development of new construction techniques, their area of activity has extended to the continental shelf and to the open ocean.

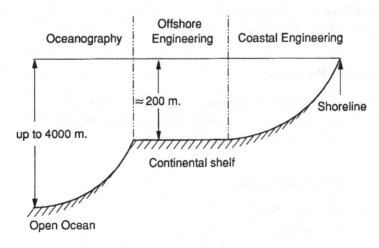

Figure 1.1 General view of engineering territory.

In order to get a picture of the activities of this field, I selected examples of current research topics in the field of coastal engineering, as shown below.

(1) Design of Coastal and Ocean Structures

a) Wave Mechanics: Wave is the major driving force that governs the design of structures. Also it is important for port and harbor construction.
b) Directional Irregular Wave
c) Wave Force on Structures
        Near-shore Current, in particular Wave Induced Current
d) Design Standard for Coastal Structures
e) New Design Technique including Landscape Planning

(2) Influence of Structures to Environment

a) Sediment Transport, Siltation (Dredging) Problem
b) Mud Transport Model
c) Local Scour, Beach Topography Change
d) Three Dimensional Modeling of Topography Change
e) Surf Zone Hydrodynamics
f) Role of Long Waves to Environment

(3) Disaster Prevention

a) Numerical Simulation of Storm Surge and Tsunami
b) Mechanism of Disaster under Storm Surge and Tsunami
c) Floods in Coastal Urban Area

(4) Environmental Problem

a) Water Quality in Bay
b) Model for Diffusion and Dispersion
c) Global Sea Level Rise and Its Effect to Local Beach Erosion

A number of books have been written in this area. For practical examples involving coastal protection, the following book is useful: U.S. Army Corps of Engineers, Coastal Engineering Manual, Part I to VI, 2006.

However, for the most updated information on the field, the reader should refer to the following journals or proceedings.

Coastal Engineering Journal (CEJ, World Scientific and JSCE),
Coastal Engineering (Elsevier),
Journal of Waterway, Port, Coastal and Ocean Engineering (ASCE),
Proceeding of Coastal Engineering Conference (ASCE),
Proceedings of Asia and Pacific Coast (APAC), and
Proceedings of Coastal and Port Engineering in Developing Countries (COPEDEC).

1.1 Three Examples of Japanese Experiences of Coastal Environment Change Due to Construction Works

The possible change in the environmental system due to coastal area developments is an important factor in planning new projects in the coastal and near-shore regions. This environmental change is very complicated and consists of individually complex processes. The response time of the coastal environment to the development project ranges from an immediate change to the scale of decades. Among these responses, the long term response is more important than the short term one, because the long term response affects a wider area and the mechanism of change is less easily understood and therefore more difficult to predict before the project starts. Three typical examples of Japanese experiences regarding coastal protection works or coastal environment works in the 150 years after the modernization process started in Japan are presented in the next sections.

1.1.1 *Ookozu channel for flood control of the Shinano River*

The Shinano River is 367 km long and it is the longest river in Japan. Before the 19th century, the main flow of the Shinano River crossed the Niigata plain. River flooding occurred frequently and seriously disrupted rice production in the Niigata plain. In the late 19th century, the Japanese Government started a project to stop flooding in the Shinano River. However, it was very hard to control the flooding, and in 1909 the government started the construction of a new channel for flood control. Figure 1.2 shows the geometry of the Shinano River and the location of the Ookozu Channel. During flood times, it is possible to divert water into the Ookozu channel and thus decrease the flow along the Shinano River.

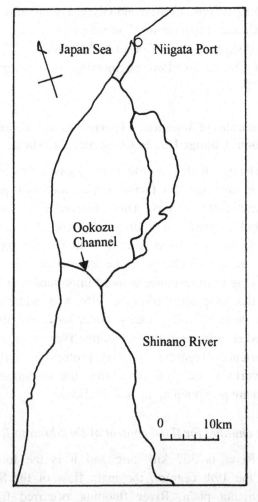

Figure 1.2 Geometry of the Shinano river and Ookozu flood channel.

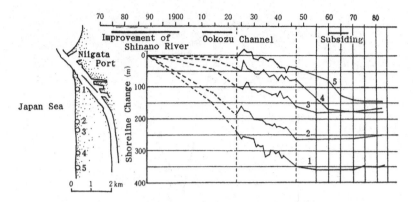

Figure 1.3 Time history of shoreline change in the vicinity of Shinano river mouth. (Data from Ministry of Land, Infrastructure and Transport.)

This channel was also good for the maintenance of Niigata port, located in the river-mouth of the Shinano River. Previously, it was difficult to maintain the waterway between the open ocean and Niigata port due to the large amount of sand discharge from the Shinano River. After the completion of Ookkozu channel, the amount of sediment discharge to original river mouth was diminished.

At the same time, erosion of Niigata coast started and continues till present. Wave attack to Niigata coast in winter season is severe, with significant wave heights of up to 5m and wave periods of up to 10s. Figure 1.3 shows the time history of shoreline change from 1890 to 1980. After the completion of Ookozu channel in 1922, the rates of erosion become quite severe, and the graph shows the erosion recorded at Mitonori beach close to the river mouth is more than 350m. In order to make the beach stable, it is necessary to continue constructing new protection works which require large sums of money and effort.

1.1.2 Reclamation work in Tokyo Bay

In 1960's, Japan experienced rapid growth in its economy. This development required areas to be reclaimed from the sea in Tokyo, Osaka and Ise Bay areas because most of the population, infrastructure and capital was concentrated in these areas. From 1960 to 1972, a total of

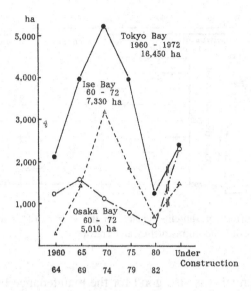

Figure 1.4 Time history of reclamation area in three major bays in Japan. (Data from Ministry of Land, Infrastructure and Transport, 1983.)

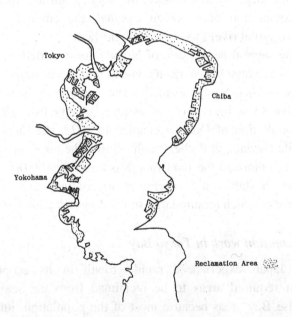

Figure 1.5 Reclamation area in Tokyo bay. (Data from Ministry of Land, Infrastructure and Transport.)

16,450 ha were reclaimed in Tokyo bay. Figure 1.4 shows the time history of reclamation in the three major bay areas in Japan. Reclamation still continues at present, with Fig. 1.5 showing the original coastline in Tokyo bay and the area that has already been reclaimed. In the reclamation of the northeast side of the bay (an area called Chiba), during the 1960's the reclamation material was taken from the sea bottom close to the reclamation area.

When the reclamation material was taken from the sea bottom, big holes were left after taking the materials. Figure 1.6 shows the topography of the area with holes. Nowadays it is not allowed to take materials from the bottom, but the holes still exist.

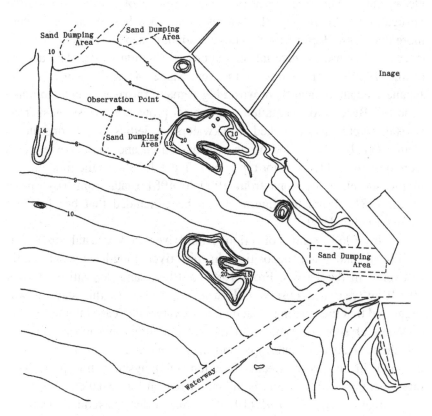

Figure 1.6 Bottom topography in Chiba area. (Data from marine chart, Marine Safety Agency.)

In these thirty years a 'blue tide', which is a water body with low or no oxygen solved, frequently appears in the Chiba area. The mechanism of the generation of blue tide was studied and it appears that a mass of water with a small concentration of oxygen forms at the bottom of the hole and then it moves under the effect of stratified flow in the summer season. Currently efforts are being made to cover the hole and make the bottom flat, but it will take a long time to recover the whole area.

1.1.3 *Coastal protection works in Suruga Bay*

The Suruga coast is located in the middle part of Japan on the Pacific Ocean side. The Abe River is a major source of beach sand for the Suruga coast. Previously, the widths of the beaches in this area were more than 70 m. Before 1968, a large amount of sand was taken from the river area for use as material for major construction works. The decade of the 1960's corresponds to the first stage of rapid growth in the Japanese economy and there was a big demand for river sand as to make concrete. Beach erosion started and coastal protection measures such as coastal revetments, detached breakwaters or jetties were constructed from 1959. In 1977, rapid beach erosion started and coastal revetments were destroyed due to major typhoons. Figure 1.7 shows the time history of the shoreline locations from 1969 to 1981 (data from Toyoshima *et al.*, 1981). From the figure, it can be concluded that beach width decreased after 1969.

In 1968, sand collection from river and coastal area was prohibited and sand deposition in the river mouth area gradually recovered in the 1970's. From 1983 to 1993, sand deposition of over $10 \times 10^4 \mathrm{m}^3$/year was observed (Uda *et al.*, 1994). At this stage it was suggested that the existing detached breakwaters decreased the longshore transport of supplied sand from the river mouth. This means that the detached breakwaters interrupt sand movement and diminish the supply of sand in the downstream direction of longshore transport. This intersects the sand flow and is the major reason for beach erosion in the downstream area. Uda *et al.* (1994) estimated that the reduction rate of sand velocity due to detached breakwaters in this area is between 71 to 53%.

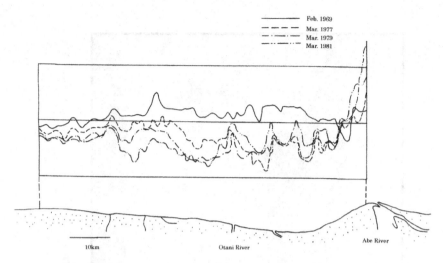

Figure 1.7 Shoreline changes in Shizuoka Coast from 1969 to 1981. (Modified from Toyoshima *et al.*, 1981.)

Three typical examples of the impact of a development project to the coastal environment in Japan were explained in the previous sections. From these three examples we can say that the long-term impact of new construction is sometimes not easily predicted. Coastal engineers should be very careful to monitor the change in the vicinity of their projects and be flexible to cope with environmental changes. Nowadays we can find the same type of mechanisms of environment change in the coastlines of many developing countries in Asia and Africa.

1.2 General View of River Sediment Supply to Coastal Area over the World (Tuan and Shibayama, 2003)

During last few decades, a number of numerical models have been developed to simulate the sediment transport process. A common aspect of these models is that they need sediment boundary conditions to accurately simulate the process. Since the major source of the sediments that are transported to the coastal zone are the rivers that drain through the land (Milliman and Meade, 1983), the boundary conditions of these models are often calculated by using observed river sediment discharge at or near the river mouths. Figure 1.8 shows major river basins over the

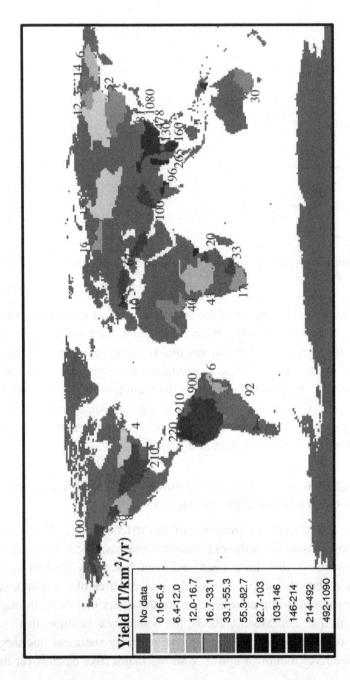

Figure 1.8 Sediment yields (ton/km²/year) and measured total sediment discharge (10^6 ton/year) of major rivers in the world. (Data from Milliman *et al.*, 1983.)

world and their total sediment dischage to the coastal environment. Big amounts of sediments are produced and are transported to the coast in major Asian rivers, such as the Mekong, Chao Phraya, Irrawadhy, Ganzes, Indus and Yangtze rivers. Also a big amount of sediment discharge is found in Indonesia. The Amazon, Mississippi and Danube are also big sources of sediment supply.

In the Asian coast, big sediment discharge from inland and coastal erosion due to high waves were balanced in previous times. But in recent years, sediment supplies from river mouths have been changing due to the rapid change of land use in their river basins. When it decreases, coastal erosion occurs and when it increases, coastal deposition appears. Rapid changes of coastal line in Asia is strongly controlled by supply of sand from the river basins.

References

Le Trung Tuan and Shibayama, T. (2003): Application of GIS to evaluate long-term variation of sediment discharge to coastal environment, *Coastal Eng. Journal*, 45(2), 275-293.

Milliman, J.D. *et al.* (1983): World-wide delivery of river sediment to Oceans, *Jour. of Geology*, 91, 1-21.

Toyoshima, O., Takahashi, W. and Suzuki, I. (1981): Characteristics of the beach erosion in Shizuoka Coast, *Proc. of Coastal Eng. Conf.*, JSCE, Vol. 28, pp. 261-265 (in Japanese).

Uda, T., Suzuki, T., Oonishi, M., Yamamoto, Y. and Itabashi, N. (1994): Evaluation of longshore Sediment transport in Suruga bay, *Proc. Coastal Eng. Conf.*, JSCE, Vol. 41, pp. 536-540 (in Japanese).

Chapter 2

Review of Fundamental Fluid Mechanics

Wave motion is the major driving force of physical phenomena in coastal areas. To understand the physical mechanism behind it requires to understand the fundamentals of fluid mechanics. A quick review of fluid mechanics shall then be presented below in order for the reader to understand the principles underlying wave mechanics or coastal hydrodynamics.

2.1 Brief History of Fluid Mechanic

Fluid motion can be divided into two categories: the first category is perfect (non-viscous) fluid and the second is viscous fluid. In the history of fluid mechanics, there were three major development stages:

1. In the middle of 18th century, Euler and Bernoulli established theories for a perfect fluid which included conservation laws for mass, momentum or kinetic energy. After their achievement, there was considerable progress in how to explain many of the empirical formulas of hydraulics from the theoretical side.

2. In the early 19th century, Navier, a French civil engineer designed a bridge base to be constructed in the River Seine. He found out that perfect fluid theory did not provide a good estimation of the flow forces on structures. He then included a viscous term to momentum conservation equation and developed a viscous flow theory (1822). Meanwhile, in England, Stokes, who was an applied mathematician, also independently developed his own viscous flow theory (1845). The momentum conservation equation for viscous flow which has been of use till today is therefore named after these two people and is

commonly referred to as Navier-Stokes equation. However, this equation is non-linear, and is not easy to use in calculating general cases. Its non-linearity has caused the viscous flow model to be limited in its area of application.

3. In the early 20th century a German researcher, Prandtl, presented a boundary layer theory (1904). The theory assumes that the flow can be divided into two layers, one of which is a perfect fluid and the other a viscous layer on the surface of structures (boundary layer). He developed a theory to analyze the boundary layer. The theory was very powerful to solve several practical problems; one example being the success in the development of the airplane as achieved by Wright brothers.

2.2 Brief Review of Vector Analysis

Vector expressions are usually used to describe flow motion. A review of the basic principles of vector analysis is therefore necessary to understand the basic principles of flow motion.

2.2.1 *Introduction of vector and scalar operators*

Vector operator "del" is defined as

$$\nabla = \vec{i}\,\frac{\partial}{\partial x} + \vec{j}\,\frac{\partial}{\partial y} + \vec{k}\,\frac{\partial}{\partial z} \tag{2.1}$$

where $\vec{i}$, $\vec{j}$, $\vec{k}$ are unit vectors of x, y and z direction, respectively. The product with scalar $\phi(x,y,z)$ is

$$grad\ \phi = \nabla\phi = \vec{i}\,\frac{\partial\phi}{\partial x} + \vec{j}\,\frac{\partial\phi}{\partial y} + \vec{k}\,\frac{\partial\phi}{\partial z}. \tag{2.2}$$

Inner product with velocity vector $\vec{u}(u,v,w)$ is

$$div\ \vec{u} = \nabla\cdot\vec{u} = \frac{\partial u}{\partial x} + \frac{\partial v}{\partial y} + \frac{\partial w}{\partial z}. \tag{2.3}$$

Outer product with vector $\vec{u}(u_x, u_y, u_z)$ is

$$rot\ \vec{u} = \nabla \times \vec{u} = \begin{bmatrix} i & j & k \\ \dfrac{\partial}{\partial x} & \dfrac{\partial}{\partial y} & \dfrac{\partial}{\partial z} \\ u & v & w \end{bmatrix}$$

$$= \vec{i}\left(\frac{\partial w}{\partial y} - \frac{\partial v}{\partial z}\right) + \vec{j}\left(\frac{\partial u}{\partial z} - \frac{\partial w}{\partial x}\right) + \vec{k}\left(\frac{\partial v}{\partial x} - \frac{\partial u}{\partial y}\right). \tag{2.4}$$

The operator Laplacian Δ is

$$\Delta \equiv \nabla \cdot \nabla = \frac{\partial^2}{\partial x^2} + \frac{\partial^2}{\partial y^2} + \frac{\partial^2}{\partial z^2} \tag{2.5}$$

$$\Delta \phi = div(grad\ \phi) = \frac{\partial^2 \phi}{\partial x^2} + \frac{\partial^2 \phi}{\partial y^2} + \frac{\partial^2 \phi}{\partial z^2} = \nabla^2 \phi. \tag{2.6}$$

The following relations are important and can be easily obtained by the above definitions.

$$div(rot\ \vec{u}) = 0$$

$$rot(grad\ \phi) = 0.$$

It is also important to visualize the physical meaning of the above operators. As an example, the meaning of rotation will be given in the next section.

2.2.2 *The physical meaning of rotation (Naganuma, 1987)*

As an example, we take the z component of rotation $\vec{u}$, that is

$$(rot\ \vec{u})_z = \frac{\partial v}{\partial x} - \frac{\partial u}{\partial y}. \tag{2.7}$$

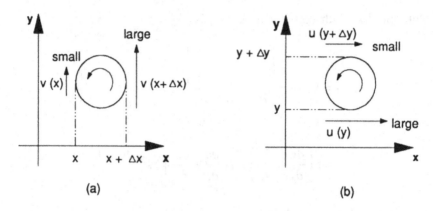

Figure 2.1 Explanation for rotation.

A disk with a radius Δx rotating counter-clockwise is considered (Figure 2.1(a)). The velocity difference between the two sections is $v(x+\Delta x)-v(x)$ and the resultant radian velocity is

$$\frac{v(x+\Delta x)-v(x)}{\Delta x}.$$

Then we take the limit to $\Delta x \to 0$

$$\lim_{\Delta x \to 0}\frac{v(x+\Delta x)-v(x)}{\Delta x}=\frac{\partial v}{\partial x}. \qquad (2.8)$$

This is the first term in right hand side. The same argument can be applied to $\frac{\partial u}{\partial y}$. Figure 2.1(b) shows us the situation

$$\lim_{\Delta y \to 0}\frac{u(y)-u(y+\Delta y)}{\Delta y}=-\frac{\partial u}{\partial y}. \qquad (2.9)$$

In total, the z component of the rotation becomes $(rot\,\vec{u})_z=\frac{\partial v}{\partial x}-\frac{\partial u}{\partial y}$. The x and y components can be obtained in the same manner but the first term is positive and the second term is negative.

2.3 Conservation Laws for Fluid Mechanics

There are generally two ways to describe the physical phenomena of fluid mechanics. The first is by Lagrangean description. In this way, we specify a fixed particle and trace this particle's movement. The location of the particle $\vec{r}_p$ is a function of its initial location and time, (x_0, y_0, z_0, t).

$$\vec{r}_p(t) = \vec{r}_p(x_0, y_0, z_0, t). \tag{2.10}$$

The second way is by Eularian description. We specify a fixed location and measure the fluid's velocity time history. Velocity vector $\vec{u}$ is a function of location and time.

$$\vec{u} = \vec{u}(x, y, z, t). \tag{2.11}$$

The relation between these two approaches is given as follows

$$\vec{u}(x, y, z, t) = \frac{d\vec{r}_p}{dt}\bigg|_{at\ time\ t} \tag{2.12}$$

and

$$\vec{r}_p(t) = \int_0^t \vec{u}(x_p, y_p, z_p, t)dt \tag{2.13}$$

where vector $\vec{r}_p$ indicates the particle's location at each time step. In the following sections, we will use the Eularian approach to describe fluid motion.

Conservations of mass, momentum and energy are the basic principles in fluid dynamics. These three conservation laws will therefore be explained in the section below.

2.3.1 *Mass conservation*

Figure 2.2 shows the inflow and outflow of mass in a unit volume. The difference between inflow and outflow results in a mass change of the unit volume. The resulting equation of this relationship is known as mass conservation equation.

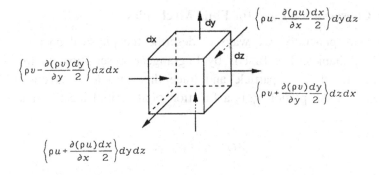

Mass change per unit volume per time

Inflow

$$\left\{\frac{\partial(\rho dxdydz)}{\partial t}\right\}dt = [\left\{\rho u - \frac{\partial(\rho u)dx}{\partial x}\frac{dx}{2}\right\}dydzdt + \left\{\rho v - \frac{\partial(\rho v)dy}{\partial y}\frac{dy}{2}\right\}dzdxdt$$

$$+ \left\{\rho w - \frac{\partial(\rho w)dz}{\partial z}\frac{dz}{2}\right\}dxdydt]$$

Outflow

$$- [\left\{\rho u + \frac{\partial(\rho u)dx}{\partial x}\frac{dx}{2}\right\}dydzdt + \left\{\rho v + \frac{\partial(\rho v)dy}{\partial y}\frac{dy}{2}\right\}dzdxdt$$

$$+ \left\{\rho w + \frac{\partial(\rho w)dz}{\partial z}\frac{dz}{2}\right\}dxdydt]$$

Figure 2.2 Mass conservation.

$$\frac{\partial\rho}{\partial t}+\frac{\partial(\rho u)}{\partial x}+\frac{\partial(\rho v)}{\partial y}+\frac{\partial(\rho w)}{\partial z}=0 \qquad (2.14)$$

where ρ is the density of the fluid, and in vector expression ($\frac{D}{Dt}\equiv\frac{\partial}{\partial t}+u\frac{\partial}{\partial x}+v\frac{\partial}{\partial y}+w\frac{\partial}{\partial z}$)

$$\frac{D\rho}{Dt}+\rho\nabla\cdot\vec{u}=0 \qquad (2.15)$$

or

$$\frac{\partial\rho}{\partial t}+\vec{u}\cdot grad\ \rho+\rho\ div\ \vec{u}=0. \qquad (2.16)$$

The first term represents the time variation of density, the second term represents the spatial distribution of density. However, these two terms are important only for density wave calculations such as the calculations for shock wave or sound wave propagation. In many cases, density ρ can be assumed to be constant for water flow.

If $\frac{D\rho}{Dt} = 0$ then Eq. (2.15) becomes

$$\nabla \cdot \vec{u} = 0. \tag{2.17}$$

Equation (2.17) is called the *continuity equation*.

2.3.2 *Momentum conservation*

The starting point is Newton's law of motion.

$$M \frac{d\vec{u}}{dt} = \vec{F} \tag{2.18}$$

where M is the mass and $\vec{F}$ is the force vector. Here, the vector $\vec{u}$ has components (u, v, w) and the x-component u is a function of (x, y, z, t).

Then

$$du = \frac{\partial u}{\partial t} dt + \frac{\partial u}{\partial x} dx + \frac{\partial u}{\partial y} dy + \frac{\partial u}{\partial z} dz. \tag{2.19}$$

Dividing by dt

$$\frac{du}{dt} = \frac{\partial u}{\partial t} + \frac{\partial u}{\partial x} \frac{dx}{dt} + \frac{\partial u}{\partial y} \frac{dy}{dt} + \frac{\partial u}{\partial z} \frac{dz}{dt}. \tag{2.20}$$

Here, $\frac{dx}{dt} = u$, $\frac{dy}{dt} = v$, $\frac{dz}{dt} = w$. Then we define operator $\frac{D}{Dt}$ as

$$\frac{Du}{Dt} \equiv \frac{\partial}{\partial t} u + u \frac{\partial u}{\partial x} + v \frac{\partial u}{\partial y} + w \frac{\partial u}{\partial z}. \tag{2.21}$$

Then, $\frac{du}{dt} = \frac{Du}{Dt}$

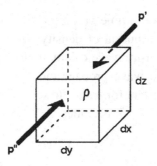

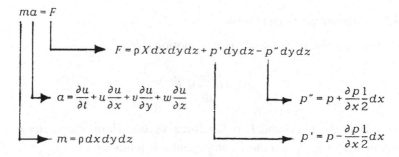

Figure 2.3 Momentum conservation without viscosity.

Figure 2.3 shows the force balance exerted on the unit mass. The momentum conservation will be (Euler Equation)

$$\rho \frac{D\vec{u}}{Dt} = \rho \vec{F} - grad\ p \qquad (2.22)$$

where p is the pressure.

For a viscous fluid, the viscous force should be added to momentum conservation. The forces to be added are drawn in Figure 2.4. For the x-component,

$$\rho \frac{Du}{Dt} = \rho x - \frac{\partial p_x}{\partial x} + \frac{\partial \tau_{yx}}{\partial y} + \frac{\partial \tau_{zx}}{\partial z}. \qquad (2.23)$$

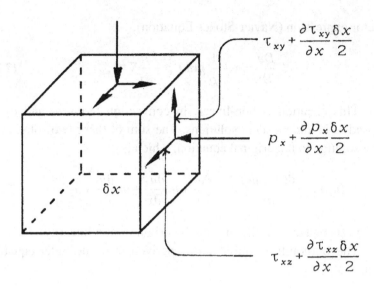

$$\rho \frac{Du}{Dt} = \rho X - \frac{\partial p_x}{\partial x} + \frac{\partial \tau_{yx}}{\partial y} + \frac{\partial \tau_{zx}}{\partial z}$$

Figure 2.4 Viscous force added to momentum conservation equation.

Here τ_{yx} is the x-directed stress in y-plane (plane perpendicular to y-axis). And

$$\tau_{yx} = \mu \left(\frac{\partial v}{\partial x} + \frac{\partial u}{\partial y} \right) \tag{2.24}$$

$$\tau_{zx} = \mu \left(\frac{\partial u}{\partial z} + \frac{\partial w}{\partial x} \right) \tag{2.25}$$

where μ is the viscosity. The normal stress can be obtained after substituting the values of τ_{yx} and τ_{zx},

$$p_x = p - 2\mu \frac{\partial u}{\partial x}. \tag{2.26}$$

In vector expression (Navier-Stokes Equation),

$$\frac{D\vec{u}}{Dt} = \vec{x} - \frac{1}{\rho} grad\ p + \frac{\mu}{\rho} \nabla^2 \vec{u} .\qquad(2.27)$$

This equation is non-linear. In convection terms u_1 and u_2 are supposed to be independent solutions. The sum of these two solutions is not the solution to the original equation, which is

$$(u_1 + u_2)\frac{\partial(u_1 + u_2)}{\partial x} = u_1 \frac{\partial u_1}{\partial x} + u_1 \frac{\partial u_2}{\partial x} + u_2 \frac{\partial u_1}{\partial x} + u_2 \frac{\partial u_2}{\partial x} .\qquad(2.28)$$

Here, the second and third terms are the origin of non-linearity.

The Bernoulli equation can be derived from the Euler equation. We put

$$q^2 = u^2 + v^2 + w^2 \qquad(2.29)$$

then

$$\frac{\partial}{\partial x}\left(\frac{1}{2}q^2\right) = u\frac{\partial u}{\partial x} + v\frac{\partial v}{\partial x} + w\frac{\partial w}{\partial x} .\qquad(2.30)$$

If we substitute this from the convection term,

$$u\frac{\partial u}{\partial x} + v\frac{\partial u}{\partial y} + w\frac{\partial u}{\partial z} - \frac{\partial}{\partial x}\left(\frac{1}{2}q^2\right) = \{w(rot\ \vec{u})_y - v(rot\ \vec{u})_z\} \qquad(2.31)$$

We substitute $\frac{\partial}{\partial x}(\frac{1}{2}q^2)$ from both sides of Euler equation and then we get a vector expression,

$$\frac{\partial\vec{u}}{\partial t} + rot\ \vec{u} \times \vec{u} = \vec{F} - \frac{1}{\rho} grad\ p - grad\left(\frac{1}{2}q^2\right).\qquad(2.32)$$

If we assume an irrotational fluid ($rot\ \vec{u} = 0$) and use $\vec{u} = grad\ \phi$,

$$\vec{F} = grad\left(\frac{\partial\phi}{\partial t} + \frac{1}{2}q^2 + \int\frac{dp}{\rho}\right).\qquad(2.33)$$

We define $\vec{F} = -grad\ \pi$ (where π is gravity potential)

$$grad\left(\frac{\partial \phi}{\partial t} + \frac{1}{2}q^2 + \int \frac{dp}{\rho} + \pi\right) = 0 \qquad (2.34)$$

Integration with space results in

$$\frac{\partial \phi}{\partial t} + \frac{1}{2}q^2 + \int \frac{dp}{\rho} + \pi = F(t). \qquad (2.35)$$

This is called generalized Bernoulli Equation. By assuming a steady state, (2.34) becomes

$$grad\left(\frac{1}{2}q^2 + \int \frac{dp}{\rho} + \pi\right) = \vec{u} \times rot\ \vec{u}. \qquad (2.36)$$

If we take the component along the stream line and insert the value for gravity potential,

$$\frac{1}{2}q^2 + \frac{p}{\rho} + gz = constant \qquad (2.37)$$

where g is acceleration of gravity.

This is Bernoulli's equation, which is used for many applications. The equation is derived from Euler's equation (momentum conservation), but the physical meaning of the equation is the conservation of dynamic energy.

2.3.3 *Energy conservation*

The energy conservation law is expressed by the state equation for a gas, which includes thermodynamical energy.

$$p = \frac{R}{m}\rho T \qquad (2.38)$$

where R is a constant ($8.314 \times 10^7\ erg/^0k$), m is molar mass and T is absolute temperature.

For fluid motion, there are five unknown variables: namely, the three components of velocity (u,v,w); pressure p and density ρ. The equations are those of mass conservation (2.15), momentum conservation, which has three equations for each direction, x, y, and z, (2.22) for inviscid and (2.27) for viscous fluid; and energy conservation (2.38). We have five equations to calculate five unknowns.

If the fluid is water, in calculating fluid motion we can assume that water is incompressible, i.e., the density ρ is constant. In that case, we are left with only four unknown variables and we can solve the problem by using the mass conservation and momentum conservation equations.

2.4 Irrotational Flow of Inviscid Fluid

Since we will be using the potential flow theory for wave analysis, a quick review will be given in this section. Now, if we go back to the 18th century where no viscosity was assumed, we could use Euler equation (2.22) for the momentum conservation equation. If we assume incompressibility, then Eq. (2.17) will be the mass conservation equation.

2.4.1 *Velocity potential*

For irrotational flow $(rot\ \vec{u} = 0)$, the velocity potential ϕ can be defined as

$$u = -\frac{\partial \phi}{\partial x}, v = -\frac{\partial \phi}{\partial y}, w = -\frac{\partial \phi}{\partial z}. \tag{2.39}$$

(It is possible to define ϕ without the minus signs as presented here. In that case, minor changes of + or − signs are required in the following parts.)

If we calculate rotation,

$$\frac{\partial u}{\partial x} - \frac{\partial u}{\partial y} = \frac{\partial}{\partial x}\left(-\frac{\partial \phi}{\partial y}\right) - \frac{\partial}{\partial y}\left(-\frac{\partial \phi}{\partial x}\right)$$

$$= 0.$$

Substituting into mass conservation equation,

$$\nabla^2 \phi = 0. \tag{2.40}$$

This is the Laplace equation and is common for flow, heat and electro-magnetic phenomena. Since this equation is a linear one, the solution can be superimposed.

2.4.2 *Stream function*

For two-dimensional flow along the stream line, the relation between line element $d\vec{s} = (dx, dy)$ and velocity $\vec{u} = (u, v)$ will be

$$\frac{dx}{u} = \frac{dy}{v}$$

and therefore

$$-udy + vdx = 0.$$

Since u and v are functions of x and y, we can define scalar ψ and

$$0 = d\psi = \frac{\partial \psi}{\partial x} dx + \frac{\partial \psi}{\partial y} dy = vdx - udy. \tag{2.41}$$

The value ψ is a stream function and the value is constant along the stream line. Since the flow is irrotational,

$$0 = \frac{\partial v}{\partial x} - \frac{\partial u}{\partial y} = \frac{\partial^2 \psi}{\partial x^2} + \frac{\partial^2 \psi}{\partial y^2}. \tag{2.42}$$

The relations between ϕ and ψ are

$$-u = \frac{\partial \phi}{\partial x} = \frac{\partial \psi}{\partial y}$$

$$-v = \frac{\partial \phi}{\partial y} = -\frac{\partial \psi}{\partial x}. \tag{2.43}$$

This is referred to as the Cauchy-Riemann relationship. Since

$$\frac{\partial \phi}{\partial x}\frac{\partial \psi}{\partial x} + \frac{\partial \phi}{\partial y}\frac{\partial \psi}{\partial y} = 0 .$$

Then the vector $(\frac{\partial \phi}{\partial x}, \frac{\partial \phi}{\partial y})$ and $(\frac{\partial \psi}{\partial x}, \frac{\partial \psi}{\partial y})$ crosses perpendicularly because the inner product is equal to zero. Tangential vectors are perpendicular; this means the lines where ϕ is equal to a constant and the lines where ψ is equal to a constant cross at right angles.

2.4.3 *Complex potential*

A complex function w is defined as

$$w \equiv \varphi + i\psi . \tag{2.44}$$

Velocity potential ϕ is the real part and stream function ψ is the imaginary part. If the condition (2.43) is correct, Cauchy-Riemann relationship, exists, w is an analytic function and

$$\frac{dw}{dz} = \frac{\partial w}{\partial x} = \frac{\partial w}{i\partial y} = -u + iv \tag{2.45}$$

since the derivative of w in terms of space does not depend on direction x or y for analytic function.

Figure 2.5 shows the examples for flow pattern with various complex potential. The examples are uniform flow, source and vortex.

Reference

Naganum, S. (1987): *Meanings of mathematical methods in physics*, pp. 49-56, Tsuusankenkyusha, Tokyo.

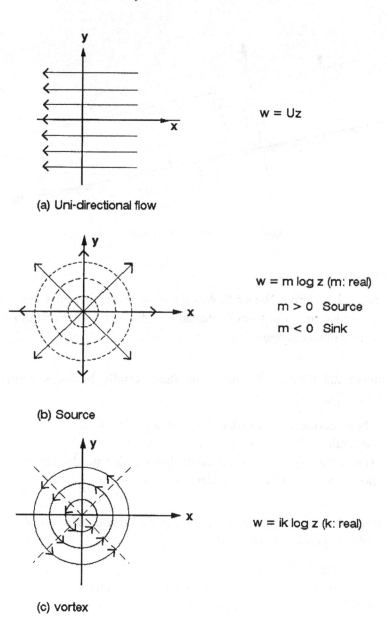

(a) Uni-directional flow

$w = Uz$

(b) Source

$w = m \log z$ (m: real)

$m > 0$ Source

$m < 0$ Sink

(c) vortex

$w = ik \log z$ (k: real)

Figure 2.5 Examples of complex potentials, w.

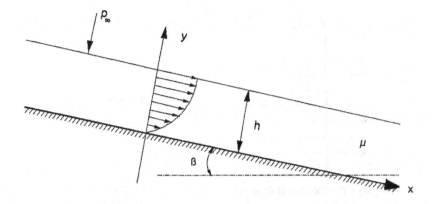

Figure 2.6 Viscous flow over sloped bed.

Exercises

Problem 2.1 Derive Navier-Stokes equation from Newton's law. Then change Euler equation from Cartesian coordinate to polar coordinate in two-dimension (Change from (x, y) to (r, θ)).

Problem 2.2 Explain the following items briefly by using formulas, figures and descriptions.

1. The importance of boundary layer theory (Prandtl).
2. Generalized Bernoulli Equation.
3. The difference between boundary layer under steady current (pipe-flow) and boundary layer under wave action.

Problem 2.3 Draw figures for velocity potential and stream function for the following complex potentials.

(a) $w = ik \log z$ (k : real and $k > 0$)
(b) $w = k \log[(z-a)/(z+a)]$ and $a \to 0$ (k : real and $k > 0$)
(c) $w = U[z \exp(-ia) + \exp(ia)/z]$, (U , a: real)
(d) $w = az^{3/4}$ (a : real)

Problem 2.4 Water flows down along a slope as shown in Figure 2.6. The flow is laminar and steady.

1. Show momentum conservation equation by neglecting unnecessary terms from Navier-Stokes Equation.
2. Show boundary conditions.
3. Use above equations and get results for the velocity (u) distribution in y-direction.

Chapter 3

Basic Equations for Wave Motion (Linear Wave Theory)

3.1 Basic Equations

The basic wave equations consist of mass conservation and several boundary conditions. There are several different wave theories depending on the different ways to solve the equation sets. The principal wave theories are the linear wave theory, Stokes wave theory and Cnoidal wave theory.

The governing equation to describe wave mechanics is the mass conservation equation. For our case, we will use Laplace's equation which will be solved with boundary conditions.

$$\nabla^2 \phi = 0 . \tag{3.1}$$

There are two basic boundary conditions: kinematic and dynamic condition. We will start with the kinematic condition. A function F is defined in the boundary

$$F(x, y, z, t) \equiv 0 . \tag{3.2}$$

The function F is extended into a series

$$F(x + \delta x, y + \delta y, z + \delta, t + \delta t)$$

$$= F(x, y, z, t) + \frac{\partial F}{\partial x} \delta x + \frac{\partial F}{\partial y} \delta y + \frac{\partial F}{\partial z} \delta z$$

$$+ \frac{\partial F}{\partial t} \delta t + O(\delta x^2, \delta y^2, \delta z^2, \delta t^2)$$

$$\left(\text{since } \frac{\delta x}{\delta t} = u \text{ etc.} \right)$$

$$= F + \left(u\frac{\partial F}{\partial x} + v\frac{\partial F}{\partial y} + w\frac{\partial F}{\partial z} + \frac{\partial F}{\partial t} \right) \delta t + O(\delta x^2, \delta y^2, \delta z^2, \delta t^2)$$

$$= F + \frac{DF}{Dt}\delta t = 0.$$

The function F is zero, and therefore

$$\frac{DF}{Dt} = 0. \tag{3.3}$$

Please see Fig. 3.1 for definitions. At the bottom $(z = -h)$

$$F = z + h(x, y) = 0. \tag{3.4}$$

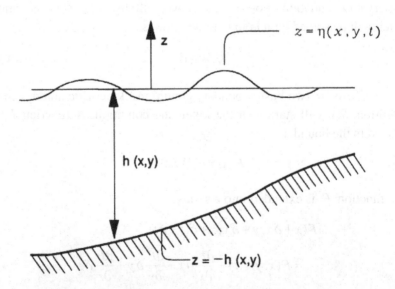

Figure 3.1 Boundary conditions.

If we substitute the functions into Eq. (3.3)

$$\frac{\partial \phi}{\partial x}\frac{\partial h}{\partial x} + \frac{\partial \phi}{\partial y}\frac{\partial h}{\partial y} + \frac{\partial \phi}{\partial z} = 0. \tag{3.5}$$

At the surface $(z = \eta)$

$$F = z - \eta(x, y, t) = 0. \tag{3.6}$$

In the same way, we get

$$\frac{\partial \phi}{\partial z} = \frac{\partial \eta}{\partial t} + \frac{\partial \phi}{\partial x}\frac{\partial \eta}{\partial x} + \frac{\partial \phi}{\partial y}\frac{\partial \eta}{\partial y}. \tag{3.7}$$

The dynamic condition is the generalized Bernoulli Equation (2.35) applied to surface. (Atmospheric pressure p_{atm} is equal to zero).

$$\frac{\partial \phi}{\partial t} + \frac{1}{2}|\vec{u}|^2 + g\eta = 0 \ (\text{at } z = \eta). \tag{3.8}$$

Equations (3.1), (3.5), (3.7), (3.8) are the set of equations used to solve wave motion. There are several different wave theories depending on the ways to solve the equation sets. The principal wave theories are the linear wave theory if we linearize the governing equation, Stokes wave theory and Cnoidal wave theory if we use perturbation technique. Figure 3.2 shows the schematic view of the difference among these wave theories.

If we use finite amplitude wave theory, it is necessary to decide whether to use Stokes wave theory or Cnoidal wave theory. In general, for the deep water condition, we use Stokes wave theory; and for shallow water condition, we use Cnoidal wave theory. For a more quantitative approach, we first calculate the Ursell parameter

$U_r (= HL^2/D^3$, H is wave height, L is wave length and D is water depth).

If U_r is smaller than, or equal to 25, we use Stokes wave theory. If U_r is greater than 25, we use Cnoidal wave theory.

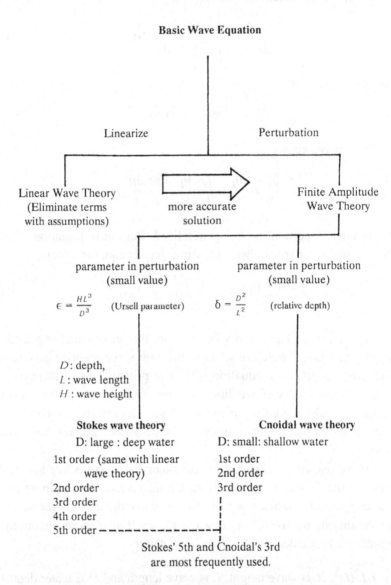

Basic Wave Equation

Linearize　　　　　　　　　　Perturbation

Linear Wave Theory
(Eliminate terms
with assumptions)　　more accurate　　Finite Amplitude
　　　　　　　　　　solution　　　　Wave Theory

parameter in perturbation　　　parameter in perturbation
(small value)　　　　　　　(small value)

$\epsilon = \dfrac{HL^3}{D^3}$　(Ursell parameter)　　$\delta = \dfrac{D^2}{L^2}$　(relative depth)

D: depth,
L: wave length
H: wave height

Stokes wave theory　　　　**Cnoidal wave theory**

D: large : deep water　　　　D: small: shallow water

1st order (same with linear　　1st order
　　　wave theory)　　　　2nd order
2nd order　　　　　　　　3rd order
3rd order
4th order
5th order

Stokes' 5th and Cnoidal's 3rd
are most frequently used.

Figure 3.2 Wave theories.

3.2 Linear Wave Theory

We will linearize the governing equations and boundary conditions by using the following three assumptions: (i) $\bar{u}$ and η are small values, (ii) two-dimensional flow and (iii) constant depth. See Fig. 3.3 for the definition of the coordinate system. The Laplace equation becomes the governing equation. From Eq. (3.1),

$$\nabla^2 \phi = 0. \tag{3.9}$$

The boundary conditions will be as follows.

From Eq. (3.5)

$$\frac{\partial \phi}{\partial z} = 0, \qquad \text{at } z = -h \qquad \text{i.e., at the bottom .} \tag{3.10}$$

From Eq. (3.7)

$$\frac{\partial \eta}{\partial t} = \frac{\partial \phi}{\partial z} \qquad \text{at } z = 0 \qquad \text{i.e., at the surface .} \tag{3.11}$$

From Eq. (3.8)

$$\frac{\partial \phi}{\partial t} + g\eta = 0 \qquad \text{at } z = 0 . \tag{3.12}$$

These four equations are the set of equations to be solved in linear wave theory.

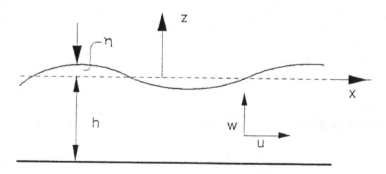

Figure 3.3 Definition for linear wave theory.

In order to solve the equation set, we will use the assumption of permanent wave, which means the wave surface profile does not change as it propagates. We will set moving coordinates with speed equal to that of wave cerelity. Thus the surface profile does not change according to time. The velocity potential for this coordinate is

$$\phi = \phi(x - ct, z). \tag{3.13}$$

Then we separate the variables

$$\phi = X(x - ct)Z(z). \tag{3.14}$$

If we insert Eq. (3.14) into Eq. (3.9), then

$$X^{"}Z + XZ^{"} = 0.$$

Since $-\frac{X^{"}}{X} = \frac{Z^{"}}{Z}$ and this is possible when it is equal to k^2 (constant), as

$$-\frac{X^{"}}{X} = \frac{Z^{"}}{Z} = k^2. \tag{3.15}$$

We shall solve X part and Z part independently. In X-part, $X^{"} + k^2 X = 0$ and the general solution is

$$X(x) = Ae^{ikx} + Be^{-ikx}. \tag{3.16}$$

Any function which is included in this general form can be a solution. Among then, we choose

$$X = \sin k(x - ct).$$

In Z-part, similarly we get

$$Z = c \cosh k(h + z). \tag{3.17}$$

In total, the solution for velocity potential is (with changing coefficient)

$$\phi = ac \frac{\cosh k(h + z)}{\sinh kh} \sin k(x - ct) \tag{3.18}$$

where c is phase velocity, k is wave number and a is an arbitrary constant, but will appears to be surface amplitude in Eq. (3.24). The value kc is called the radian frequency and we can say $kc = \sigma$.

Now, various wave characteristics can be derived from Eq. (3.18). Taking the derivative of (3.12) in terms of t,

$$\frac{1}{g}\frac{\partial^2 \phi}{\partial t^2} + \frac{\partial \eta}{\partial t} = 0. \tag{3.19}$$

Inserting (3.19) into (3.11), we obtain,

$$\frac{\partial \phi}{\partial z} = -\frac{1}{g}\frac{\partial^2 \phi}{\partial t^2} \qquad (\text{at } z = 0). \tag{3.20}$$

By using Eq. (3.18) and calculating the left-hand side and right-hand side of Eq. (3.20), we get

$$k \sinh kh - \frac{(kc)^2}{g}\cosh kh = 0. \tag{3.21}$$

By using σ,

$$\sigma^2 = gk \tanh kh. \tag{3.22}$$

This relation shows us that $c(=\sigma/k)$ changes according to k (function of T) and h, and therefore the wave phase velocity c changes according to period and depth. This means that several waves with different wave period exist at one location at the same time, and they disperse afterwards because of the difference in speed between them. The relation shown in Eq. (3.22) is frequently called the dispersion relationship.

From Eq. (3.12) and inserting Eq. (3.18)

$$\eta = -\frac{1}{g}\frac{\partial \phi}{\partial t}$$

$$= a\frac{kc^2}{g}\frac{\cosh kh}{\sinh kh}\cos k(x - ct). \tag{3.23}$$

Since $\frac{kc^2}{g}\frac{\cosh kh}{\sinh kh}$ is equal to 1 from Eq. (3.22),

$$\eta = a \cos k(x-ct). \tag{3.24}$$

Through these we have verified that the value of a is the wave amplitude. We can get the velocity field from Eq. (3.18)

$$u = \frac{\partial \phi}{\partial x} = kac\frac{\cosh k(h+z)}{\sinh kh}\cos k(x-ct) \tag{3.25}$$

$$w = \frac{\partial \phi}{\partial z} = kac\frac{\sinh k(h+z)}{\sinh kh}\sin k(x-ct). \tag{3.26}$$

The pressure is calculated from generalized Bernoulli equation,

$$p = -\rho\frac{\partial \phi}{\partial t} - \rho gz - \frac{\rho}{2}\left\{\left(\frac{\partial \phi}{\partial x}\right)^2 + \left(\frac{\partial \phi}{\partial z}\right)^2\right\} \text{ and after some calculation,}$$

$$\frac{p}{\rho g} = a\frac{\cosh k(h+z)}{\cosh kh}\cos k(x-ct) - z. \tag{3.27}$$

Here is a reminder of the relations between parameters. In Eq. (3.18),

$$kL = 2\pi \text{ therefore } k = \frac{2\pi}{L} : \text{ wave number}$$

$$kcT = \sigma T = 2\pi \text{ therefore } \sigma = \frac{2\pi}{T} : \text{ radian frequency}$$

and $c = \frac{\sigma}{k}$ therefore $\sigma = kc$.

Wave energy is the sum of kinetic energy and potential energy. These two forms of energy can be evaluated from their definition.

kinetic energy E_k :

$$E_k = \frac{1}{L}\int_0^L \int_{-h}^0 \frac{\rho}{2}(u^2 + w^2)\,dzdx$$

$$= \frac{\rho g}{4}a^2 \tag{3.28}$$

potential energy E_p :

$$E_p = \frac{1}{L}\int_0^L \left[\int_{-h}^\eta \rho g z\,dz - \int_{-h}^0 \rho g z\,dz\right]dx$$

$$= \frac{\rho g}{4}a^2 . \tag{3.29}$$

The total energy is the sum of these two, namely

$$E = E_k + E_p = \frac{\rho g}{2}a^2 \tag{3.30}$$

Energy propagation is also an important phenomenon. The rate of energy flux W can be evaluated from the product of flow velocity and energy. This is

$$W = \frac{1}{T}\int_0^T \int_{-h}^\eta \vec{u}\left\{p + \frac{\rho}{2}(|u|^2 + |w|^2) + \rho g z\right\}dzdt$$

Since $\vec{u} = \frac{\partial\phi}{\partial x}$ and $\{p + \frac{\rho}{2}(|u|^2 + |w|^2) + \rho g z\} = \rho\frac{\partial\phi}{\partial t}$ (from generalized Bernoulli Equation),

$$W = \frac{1}{T}\int_0^T \int_{-h}^\eta -\rho\frac{\partial\phi}{\partial x}\frac{\partial\phi}{\partial t}\,dzdt$$

$$= Enc . \tag{3.31}$$

Here, $n = \frac{1}{2}(1 + \frac{2kh}{\sinh kh})$ and $\frac{1}{2} \le n \le 1$. We define group velocity c_G as,

$$c_G = nc . \tag{3.32}$$

The physical meaning of c_G is energy transfer velocity. The value c_G is equal to the phase velocity c only in very shallow water area where $(kh \rightarrow 0)$. In other cases c_G is smaller than c. This gives us a somewhat strange feeling that the individual wave propagation is much faster than the transport of wave energy. For a wave group transformation, the top runner of the group gradually disappears and just behind the wave group, a new wave is produced. This means that the total speed of wave group transformation is smaller than the individual wave phase velocity.

Exercises

Problem 3.1 A wave period of 6 s. height H = 1 m, is recorded by a bottom pressure type wave gage in 15 m depth in water. Assume the wave to be sinusoidal and that the linear wave theory applies.

(1) What is the length of this wave?
(2) What is the wave celerity of this wave?
(3) What is the pressure variation measured by the pressure gage?
(4) What is the amplitude of the horizontal and vertical velocity at the bottom?
(5) What is the total energy of this wave?

(Modified from the homework problem of Prof. Madsen in 1977.)

Problem 3.2 The two velocity potentials are given.

$$\phi_1 = ac\frac{\cosh k(h+z)}{\sinh kh}\sin k(x-ct)$$

$$\phi_2 = -ac\frac{\cosh k(h+z)}{\sinh kh}\sin k(x+ct)$$

Assume linear wave theory to be valid. (This problem is for standing wave).

(1) Describe what ϕ_1 and ϕ_2 means from a physical point of view? Give emphasis to wave direction.

(2) Define a velocity potential $\phi = \phi_1 + \phi_2$ and show that $\frac{\partial \phi}{\partial x} = 0$ at $x = 0$. Would the motion described by ϕ be changed if a vertical impermeable wall was introduced at $x = 0$? Explain the physical meaning of this phenomena.

(3) Assuming the linear theory to be valid, determine the surface profile corresponding to ϕ and sketch it in the interval $0 \le x \le L$ at times corresponding to $\sigma t = 0, \frac{\pi}{2}, \pi, \frac{3\pi}{2}$.

(Modified from the homework problem of Prof. Madsen in 1977.)

Problem 3.3 A wave maker generates two groups of waves in a long wave flume. The first group has a wave period of 1.2 s, the second group has a period of 1.4 s. The water depth in the flume is 1.5 m. After generation of the 1.2 s waves, the generator is stopped for 10 s before the 1.4 s waves are generated. How far down the flume will be the front of the 1.4 s waves when it catches up with the 1.2 s waves?

(Modified from the homework problem of Prof. Madsen in 1977.)

Problem 3.4 A wave of period 8 s and deep water wave height 0.4 m, is coming to a beach.

(1) Determine wave height and wave length at the point where the water depth is 2 m.
(2) Calculate the ratio between energy velocity (group velocity) and wave phase velocity (water depth: 2 m).
(3) When you use finite amplitude wave theory, which wave theory is appropriate to use, the cnoidal wave theory or the Stokes wave theory? (water depth: 2 m)

Problem 3.5 The surface profile η of a second order Stokes wave is given as

$$\eta = a \cos k(x - ct) + a \frac{ka \cosh kh (3 + 2\sinh^2 kh)}{4 \quad \sinh^3 kh} \cos 2k(x - ct).$$

Simplify the solution of the surface profile of a second order Stokes wave, using the assumption of long waves, to obtain a simple expression for the amplitude of the second harmonic relative to the amplitude of the first harmonic. Show that this ratio is a function of the Ursell parameter. (The assumption of long waves: $kh \to 0$, $\cosh kh \to 1$ and $\sinh kh \to kh$.)

As the amplitude of the second harmonic increases relative to the amplitude of the first harmonic, the possibility of a profile with a small secondary crest in the trough of the primary wave arises. Deduce analytically the maximum value of the Ursell parameter for which the profile does not exhibit secondary crest.

(Modified from the homework problem of Prof. Madsen in 1977.)

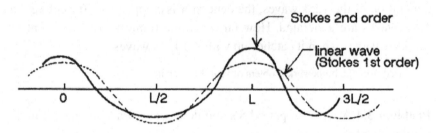

Figure 3.4 Appearance of secondary crest.

Problem 3.6 Determine which waves, Stokes wave or Cnoidal wave, should be used in the following conditions. (Use linear wave theory to calculate Ursell parameter)

(1) Wave height: 2 m   (2) Wave height: 2 m   (3) Wave height: 2 m
     Wave period: 7 s   Wave period: 7 s   Wave period: 4 s
     Water depth: 15 m   Water depth: 5 m   Water depth: 5 m

(Note: The choice of transition from one wave theory to the other theory depends on the individual problem. It is left up to the individual. Problem 3.5 gives only a first approximation).

Chapter 4

Wave Induced Physical Phenomena

Wave movements in coastal and nearshore waters are manifested in various physical consequences. Among these wave induced effects are two essential physical phenomena that are of big interest to coastal engineers. In this chapter, therefore, we shall be limiting our discussions to the mass transport velocity and bottom boundary layer which characterizes the basic theory of wave mechanics.

4.1 Mass Transport Velocity

Wave motion is formulated by using the Eularian description in Chapter 3. If we use the Lagrangean description, each water particle is transported in the wave propagation direction under wave action even for linear wave theory. The transport of water particle in wave direction is called the Lagrangean mass transport velocity.

Particle orbits of water particle can be calculated by using the equations described below. The location of each water particle (x, z) can be expressed as the sum of the mean location $(\bar{x}, \bar{z})$ and deviation (ξ, ζ) from there, as

$$x(t) = \bar{x} + \xi(t), \quad z(t) = \bar{z} + \zeta(t).$$

The derivative in terms of time is

$$\frac{dx}{dt} = \frac{d\xi}{dt} = u\,(\bar{x} + \xi, \bar{z} + \zeta, t)$$

$$= u(\bar{x}, \bar{z}, t) + \xi \frac{du}{dx} + \zeta \frac{du}{dz} + \cdots \qquad (4.1)$$

43

The first term gives a first order solution, while the second and third terms give a second order solution. If we take only the first order, and integrate it with time, the particle orbits for the first order solution become

$$\xi = -a\frac{\cosh k(\bar{z}+h)}{\sinh kh}\sin(k\bar{x}-\sigma t) \qquad (4.2)$$

$$\zeta = a\frac{\sinh k(\bar{z}+h)}{\sinh kh}\cos(k\bar{x}-\sigma t). \qquad (4.3)$$

These equations give a closed orbit, and hence a water particle will return to the starting point after one wave period. This means that water is not transported in the wave propagation direction.

If we take the second order and take the time average to get the mean mass transport U averaged over one wave period, that is;

$$U = \overline{u(\bar{x},\bar{z},t)} + \overline{\left(\int_0^t u\,dt\right)\frac{\partial u}{\partial x}} + \overline{\left(\int_0^t w\,dt\right)\frac{\partial u}{\partial z}}. \qquad (4.4)$$

The first term is called Eularian transport U_E and the second and third terms are called Lagrangean transport U_L. By using linear wave theory

$$U_L = \frac{a^2\sigma k}{2\sinh^2 kh}\cosh 2k(\bar{z}+h) \qquad (4.5)$$

$$W_L = 0 \qquad (4.6)$$

This transport U_L is frequently called Stokes drift.

Concerning U_E, Longuet-Higgins (1953) took the effect of viscosity into consideration. He considered the existence of a laminar boundary layer at the free surface and at the bottom. Thus he obtained.

$$U_E = \frac{3k\sigma a^2}{4\sinh^2 kh} \qquad z = -h \qquad (4.7)$$

$$\frac{dU_E}{dz} = -2k^2\sigma a^2 \coth kh \qquad z = 0. \qquad (4.8)$$

One more condition is necessary to determine the vertical distribution of mass transport. This leads us to the arbitrariness of mass transport velocity.

If we assume that there is no total mass transport at all, which corresponds to a limited length in longitudinal direction and a non-permeable solid boundary at the end, then,

$$\int_{-h}^{0} (U_E + U_L)\, dz = 0. \qquad (4.9)$$

We will have the following expressions,

$$U = U_E + U_L$$

$$= \frac{ka^2\sigma}{4\sinh^2 kh}\{3 + kh(3\mu^2 + 4\mu + 1)\sinh 2kh$$

$$+ 3\left(\frac{\sinh 2kh}{2kh} + \frac{3}{2}\right)(\mu^2 - 1) + 2\cosh[2kh(1+\mu)]\} \qquad (4.10)$$

where $\mu = z/h$.

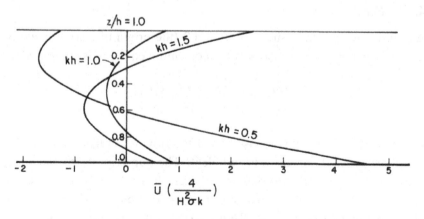

Figure 4.1 Vertical distribution of non-dimensional mass transport velocity in a closed channel (after Longuet-Higgins, 1953).

Figure 4.1 shows the vertical distribution of mass transport velocity based on the assumption (4.9). When the relative water depth h/L is small, the mass of water moves offshore at the upper layer, and moves onshore at the lower layer. When the relative water depth is great, at the upper and lower layers, the mass moves onshore, while in the middle layer, it moves offshore.

If we assume a no-viscosity condition, we have

$$U = \frac{ka^2\sigma}{2\sinh^2 kh}\cosh 2k(h+z) \qquad (4.11)$$

and the total mass transport for the condition is

$$\int_{-h}^{0} \rho U(z)dz = \frac{\rho\sigma a^2}{2}\coth kh. \qquad (4.12)$$

4.2 The Bottom Boundary Layer

Sediment motion or wave attenuation due to bottom friction is one of the most important consequences of wave motion. A careful understanding of this phenomena, however, entails a detailed consideration of the nature of the bottom boundary layer due to the oscillatory fluid motion.

4.2.1 *Formulation*

A coordinate system is shown in Fig. 4.2 (the origin of z-co-ordinate is taken at the bottom.). The horizontal momentum equation is simplified to

$$\frac{Du}{Dt} = -\frac{1}{\rho}\frac{\partial p_1}{\partial x} + \frac{\partial}{\partial z}\left(v\frac{\partial u}{\partial z}\right) \qquad (4.13)$$

in which v is the kinematic viscosity if the flow is laminar, or $v = \varepsilon$ the eddy viscosity, if the flow is turbulent. The subscript 1 indicates the values at the outer edge of bottom boundary layer. With boundary layer approximation, such as 1) change of u in x-direction is smaller than that of z-direction and 2) w is zero, the linearized boundary equation is,

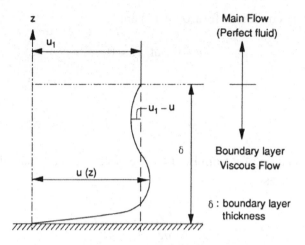

Figure 4.2 Definition sketch for boundary layer.

$$\frac{\partial u}{\partial t} = -\frac{1}{\rho}\frac{\partial p_1}{\partial x} + \frac{\partial}{\partial z}\left(\nu\frac{\partial u}{\partial z}\right) \qquad (4.14)$$

which is an equation involving only the horizontal velocity, u. Once solved, the vertical velocity, w, may be found from the mass continuity equation. At $z \cong \delta$ (at the edge of boundary layer), Eq. (4.14) then becomes

$$\frac{\partial u_1}{\partial t} = -\frac{1}{\rho}\frac{\partial p_1}{\partial x}. \qquad (4.15)$$

Since $\delta \ll h$, we may take with sufficient accuracy,

$$u_1 = u_b \cos(kx - \sigma t) \qquad (4.16)$$

where u_b is velocity amplitude at the edge of boundary layer and is given by, for example, linear wave theory or finite amplitude wave theory. If we subtract Eq. (4.15) from Eq. (4.14)

$$\frac{\partial(u - u_1)}{\partial t} = \frac{\partial}{\partial z}\left(\nu\frac{\partial(u - u_1)}{\partial z}\right). \qquad (4.17)$$

The boundary conditions for Eq. (4.17) are

$$u = 0 \quad \text{at} \quad z = 0 \tag{4.18}$$

and

$$u \to u_1 \quad \text{at} \quad z \cong \delta. \tag{4.19}$$

The determination of u from Eq. (4.17) enables us to evaluate the bottom shear stress, τ_b as

$$\tau_b = \rho v \frac{\partial u}{\partial z} \quad \text{at} \quad z = 0 \tag{4.20}$$

which is an important quantity to describe sediment transport. This is also connected with the attenuation of waves due to friction losses at the bottom.

4.2.2 *Solution of the linearized boundary layer equation*

If we assume that the solution is a periodic function, then we can introduce

$$u - u_1 = U(z) e^{i(kx - \sigma t)} \tag{4.21}$$

into Eq. (4.17), after which the equation becomes an ordinary differential equation.

$$\frac{d}{dz}\left(v\frac{dU}{dz}\right) + i\sigma U = 0 \tag{4.22}$$

with boundary conditions

$$U = -u_b \quad \text{at} \quad z = 0 \tag{4.23}$$

$$U = 0 \quad \text{at} \quad z \to \delta. \tag{4.24}$$

For laminar flow, $v = $ constant, and

$$\frac{d^2 U}{dz^2} + i\frac{\sigma}{v}U = 0.$$ (4.25)

The solution is

$$U = -u_b \exp\left\{-(1-i)\sqrt{\frac{\sigma}{2v}}z\right\}.$$ (4.26)

The velocity profile within the boundary layer is

$$u = u_b \left\{\cos(kx - \sigma t) - \exp\left(-\sqrt{\frac{\sigma}{2v}}z\right)\cos\left(kx - \sigma t + \sqrt{\frac{\sigma}{2v}}z\right)\right\}.$$ (4.27)

For turbulent flow, the treatment of Kajiura (1964) will be adopted. We then make an assumption about the form of the vertical distribution of turbulent eddy viscosity, namely,

$$v = \varepsilon = \kappa \bar{u}_* z$$ (4.28)

in which κ is Karman's constant and is equal to 0.4, and $\bar{u}_*$ is an average value of the shear velocity. If we introduce Eq. (4.28) into Eq. (4.17) with the expression of velocity deficit,

$$u - u_1 = U(z)e^{i(\sigma t - kx)}$$ (4.29)

which results in an ordinary differential equation

$$\frac{d}{dz}\left(\kappa \bar{u}_* z \frac{dU}{dz}\right) + i^3 \sigma U = 0.$$ (4.30)

The boundary conditions are

$$U = -u_b \quad \text{at} \quad z = z_0$$ (4.31)

$$U = 0 \quad \text{at} \quad z \to \delta$$ (4.32)

where z_0 is a characteristic dimension of the boundary roughness.

4.2.3 *The bottom shear stress*

For laminar flow, the bottom shear stress is given by

$$\tau_b = \rho v \frac{\partial u}{\partial z}$$

$$= \rho \sqrt{v\sigma}\, u_b \cos\left(kx - \sigma t - \frac{\pi}{4}\right). \tag{4.33}$$

Here, we notice that the maximum shear stress, τ_{bm} is given as

$$\tau_{bm} = \rho \sqrt{v\sigma}\, u_b \tag{4.34}$$

which leads the maximum velocity by a phase angle $\frac{\pi}{4}$.

Following Jonsson (1966), we use the following expression:

$$\tau_{bm} = \frac{1}{2} f_w \rho\, u_b^2 . \tag{4.35}$$

For the laminar boundary layer, if we compare Eqs. (4.34) and (4.35),

$$f_w = 2 \frac{\sqrt{v\sigma}}{u_b} = \frac{2}{\sqrt{RE}} . \tag{4.36}$$

Here, RE is the wave Reynolds number,

$$RE = \frac{u_b A_b}{v} \tag{4.37}$$

in which A_b is the horizontal amplitude of the near-bottom particle orbits.

Jonsson (1966) suggested that the critical wave Reynolds number, RE_c, for transition to the turbulent boundary layer is,

$$RE_c = 1.26 \times 10^4 \tag{4.38}$$

and a critical condition involving the Nikuradse equivalent sand roughness, k_s of the bottom is

$$\left(\frac{A_b}{k_s}\right)_c = \frac{4\sqrt{2}}{\pi}\sqrt{RE} \qquad (4.39)$$

where $(\frac{A_b}{k_s}) > (\frac{A_b}{k_s})_c$, the boundary layer is laminar if $RE < RE_c$.

Jonsson (1966) obtained the following expression for full turbulent flow:

$$\frac{1}{4\sqrt{f_w}} + \log_{10}\frac{1}{4\sqrt{f_w}} = \log_{10}\frac{A_b}{k_s} - 0.12. \qquad (4.40)$$

He presented a general wave friction factor diagram which is shown in Fig. 4.3. As shown in this figure, the wave friction factor is a function of the wave Reynolds number and the relative roughness A_b/k_s.

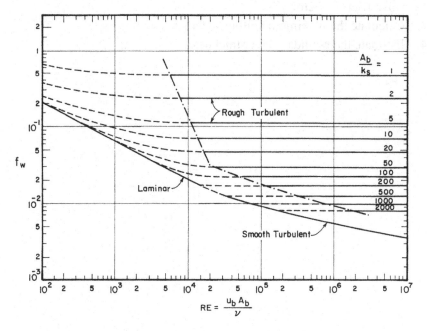

Figure 4.3 Jonsson's wave friction factor diagram (after Jonsson 1966).

References

Jonsson, I.G. (1966): Wave boundary layers and friction factors, *Proc. 10th Coastal Eng. Conf.*, pp. 127-148.

Kajiura, K. (1964): On the bottom friction in an oscillatory current, *Bull. of Earthquake Res. Inst.*, Vol. 42, pp. 147-174.

Longuet-Higgins, M.S.: *"Mass transport in water waves"*, Phil. Trans. Proc. Royal Soc. of London, Series A, p. 535-581, 1953.

Exercise

Problem 4.1 Determine the maximum bottom shear stress and the boundary layer thickness associated with a 10 s wave in a water depth of 6 m and a wave height of 1.2 m for a rough bottom of equivalent sand roughness $k_S = 2.5$ cm.

1. Calculate the values of the maximum orbital velocity at the bottom, u_b, and the excursion amplitude, A_b, at the bottom.
2. Determine the wave Reynolds number, relative roughness and the wave friction factor, f_w.
3. Calculate the maximum bottom shear stress.
4. Estimate the boundary layer thickness.

Chapter 5

Wave Transformation

In coastal areas, waves come from the offshore area and propagate into shallow waters. Due to the effect of the coastal bottom on water motion, waves are deformed according to the variations of the nearshore topography. There are several types of wave deformations, namely, shoaling, breaking, refraction, diffraction and reflection. In this chapter, we will start with descriptions of the physical meaning of wave deformation and the means to predict these deformations. At the last stage, a method to predict them using a numerical calculation will be explained.

5.1 Wave Shoaling

As waves approach the shallow water region they suffer the effect of the coastal sea bed. The energy transfer velocity of waves changes as illustrated in Eq. (3.31). This illustration gives us a rough idea of shoaling. If phase velocity c decreases, the wave length decreases, since the wave period does not change as it propagates. If the energy velocity c_g decreases and if the energy flux is constant, the energy density per unit surface area increases. This results in a transformation of the wave as it approaches the coast line, gradually increasing in height as the water depth reduces.

Generally, there are two ways to estimate wave height change due to shoaling. The first is by using the energy flux method and the second is to solve the wave equation over a sloping bottom, e.g., by mild slope equation. The mild slope equation will be described in the later part of this chapter (Section 5.6). In the energy flux method, the energy flux, i.e., the product of energy density and energy velocity, is assumed to be constant over a sloping bottom. If there is a certain amount of energy loss caused by bottom friction or wave breaking, this method cannot be applied. The energy flux is given by

$$W = \frac{1}{T}\int_0^T \int_{-h}^{\eta} u\left\{ p + \frac{1}{2}\rho(u^2 + w^2) + \rho gz \right\} dzdt \tag{5.1}$$

where u is the energy transfer velocity and the equation within the braces is the energy per unit volume. If we use Eq. (2.35), Eq. (5.1) becomes

$$W = \frac{1}{T}\int_0^T \int_{-h}^{\eta} -\rho \frac{\partial\phi}{\partial x}\frac{\partial\phi}{\partial t} dzdt. \tag{5.2}$$

Since we assume that energy flux is constant, we will use the permanent wave assumption of Eq. (3.13) where

$$\frac{\partial\phi}{\partial t} = -c\frac{\partial\phi}{\partial x}.$$

Then

$$W = \frac{1}{T}\int_0^T \int_{-h}^{\eta} -\rho c\left(\frac{\partial\phi}{\partial x}\right)^2 dzdt$$

$$= \rho c\int_{-h}^{\eta} u^2 dz. \tag{5.3}$$

For linear wave theory, Eq. (3.25) is used for the value of u, then we obtain

$$W = \frac{\rho g H^2}{8}nc = \frac{\rho g H_0^2}{8}\cdot\frac{c_0}{2} \tag{5.4}$$

where $n = \frac{1}{2}(1 + \frac{2kh}{\sinh 2kh})$ and subscript 0 means values in deep water condition.

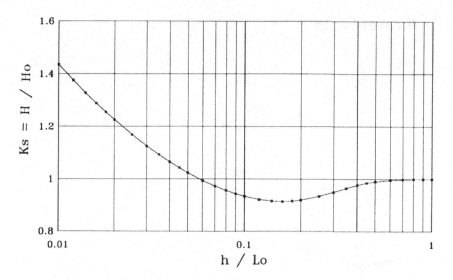

Figure 5.1 Shoaling coefficient.

The shoaling coefficient K_s is

$$K_s \equiv \frac{H}{H_0} = \sqrt{\frac{c_0}{2nc}}.$$ (5.5)

Figure 5.1 shows the value of K_s in terms of h/L_0. We may notice that as waves come towards the shallow water area from the deep water area, the wave height decreases first and then increases rapidly. The rapid increase results in depth-controlled wave breaking.

5.2 Wave Breaking

As water depth decreases, wave height increases. When the ratio between wave height and water depth becomes roughly more than 0.8, the wave breaks. This phenomenon is called depth-controlled wave breaking.

Breaker Types

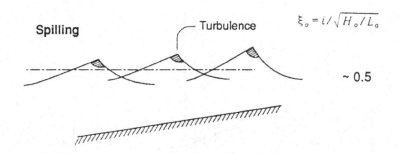

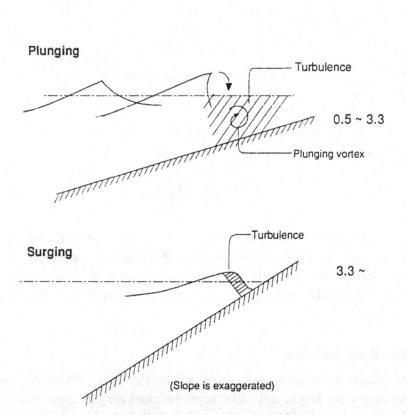

Figure 5.2 Breaker types.

The most useful formula to predict the location of wave breaking is Goda's breaking index which is based on a semi-theoretical-empirical analysis. The formula is (Goda, 1975)

$$\frac{H_b}{L_0} = 0.17\{1 - \exp[-1.5\pi h_b/L_0(1+15i^{4/3})]\} \tag{5.6}$$

where H_b: breaking wave height, h_b: breaking wave depth, i: bottom slope and L_0: deep water length.

There are three types of breakers: spilling breaker, plunging breaker and surging breaker (see Fig. 5.2). The breaker type classification is necessary from the view of energy dissipation process. The spilling breaker occurs under a gentle slope and high wave steepness. At the wave crest, the profile curls over and the energy dissipation occurs gradually over a long distance.

The plunging vortex occurs under a steeper slope and lower wave steepness. The crest plunges into the trough and gives an impression of violence to observers. The plunging vortex is formed in the vicinity of a breaking point and energy dissipation is concentrated in this area. From the view point of sediment transport or mixing process in the surf zone, the plunging breaker is more significant than the other two types of breakers due to the agitation originating from the plunging vortex.

The surging breaker occurs under a much steeper slope and a much lower wave steepness as compared to the conditions of the other two breakers. Wave breaking occurs in the vicinity of the shoreline and wave reflection is rather more important than energy dissipation.

These three types can be classified by using Irribalen parameter, (Battjes, 1974).

$$\xi_0 \frac{i}{\sqrt{H_0/L_0}}. \tag{5.7}$$

Spilling breakers occur when the parameter is less than 0.5, plunging breaker when it is between 0.5 to 3.3 and surging breaker if greater than 3.3.

5.3 Wave Reflection and Transmission

If there is a boundary in a wave field, wave reflection will occur at this boundary. The definition of reflection coefficient K_R is given as

$$K_R = \frac{a_2}{a_1} \tag{5.8}$$

where a_1 is the amplitude of the incident wave and a_2 is the amplitude of the reflected wave. Under the existence of both incident and reflected waves, a standing wave field is formed.

The value K_R changes according to the dimension and nature of the boundary. For example, in the case of a solid straight boundary, K_R is equal to one if the boundary is vertical and K_R is roughly equal to 0.1 if the boundary has an angle of 10 degrees to the horizontal plane. The value of K_R becomes smaller if the slope becomes less steep mainly due to the effect of wave breaking. If the boundary is permeable, the reflection coefficient reduces considerably.

If the top of the boundary is underwater, as in the case of submerged breakwaters, a part of the wave energy is transmitted through the boundary. In this case, the total incident energy flux is divided into three parts, transmitted, reflected and loss at the boundary.

5.4 Wave Refraction

Up to this point, we have considered the different types of wave transformations such as shoaling, breaking, reflection and transmission. All these phenomena are treated as an on-offshore direction problem (one-dimensional problem). However, there is another type of wave transformation which considers both the on-offshore and longshore directions. This consideration is important in the analysis of wave transformation and is referred to as wave refraction.

Refraction of waves occurs when waves travelling towards the shallow water area are propagating at some angle θ to the shoreline. Such a phenomenon causes the wave crest line to change its alignment as it moves over shallow waters. The change in wave crest line is brought about by the change in wave phase velocity which is greatly determined by the change in water depth. Since a one-dimensional problem will be extended to a two-dimensional one, the surface profile expressed in Eq. (3.24) can be replaced by

$$\eta = a \cos\{kx \cos\theta + ky \sin\theta - \sigma t\}. \qquad (5.9)$$

If we define the wave number vector $\vec{k} = (k\cos\theta, k\sin\theta) = (k_x, k_y)$, Eq. (5.9) becomes (see Fig. 5.3)

$$\eta = a \cos\{k_x x + k_y y - \sigma t\}. \qquad (5.10)$$

Here, the wave number potential Γ is defined as

$$\Gamma(x, y, t) = k_x x + k_y y - \sigma. \qquad (5.11)$$

The following are true for

$$\vec{k} = grad\ \Gamma \qquad (5.12)$$

and

$$k_x = \frac{\partial\Gamma}{\partial x}, \quad k_y = \frac{\partial\Gamma}{\partial y}, \quad \sigma = -\frac{\partial\Gamma}{\partial t}. \qquad (5.13)$$

From vector analysis and since Γ is a scalar,

$$rot(grad\ \Gamma) = 0. \qquad (5.14)$$

After this calculation, we can determine that

$$\frac{\partial k_y}{\partial x} - \frac{\partial k_x}{\partial y} = 0. \qquad (5.15)$$

This condition is called the irrotational condition for wave number. Also from vector analysis,

$$\frac{\partial}{\partial t}(grad\ \Gamma) - grad\left(\frac{\partial\Gamma}{\partial t}\right) = 0. \qquad (5.16)$$

After this calculation, we can determine that

$$\frac{\partial \vec{k}}{\partial t} + grad\ \sigma = 0. \tag{5.17}$$

This is known as the conservation of wave number. The set of these two equations, Eqs. (5.15) and (5.17) is the key to solve refraction problems.

We will solve this set for a simplified case. Assume a steady state and a one-dimensional wave propagation in the θ-direction (see Fig. 5.4). Since the phenomenon is uniform in the y-direction and also in time, Eq. (5.17) leads to

$$grad\ \sigma = 0 \tag{5.18}$$

therefore, the value k_y is constant. From Eq. (5.15)

$$\frac{\partial k_y}{\partial x} = 0 \tag{5.19}$$

therefore, the value of σ is constant. Here

$$k_y = k \sin \theta = \frac{\sigma}{c} \sin \theta. \tag{5.20}$$

If we substitute the value of σ (which is equal to constant) into Eq. (5.20), then we have

$$\frac{\sin \theta}{c} = \text{constant}. \tag{5.21}$$

This is called Snell's law. By using this formula, the wave angle θ can be calculated from one location to the other. The value of the phase velocity c can be determined from velocity equations such as Eq. (3.22).

The wave height change can be calculated by using the energy flux method. Figure 5.5 illustrates how the energy flux per unit length in the longshore direction can be determined. It also shows that energy flux in Section I (offshore condition with, values indicated by subscript 0) and Section II can be compared. The flux per unit of longshore length l is

$$E_0 C_{g0} l \cos \theta_0 = E C_g l \cos \theta . \tag{5.22}$$

Since $E = \frac{\rho g H^2}{8}$, the wave height ratio becomes

$$\frac{H}{H_0} = \sqrt{\frac{C_{g0}}{C_g}} \sqrt{\frac{\cos \theta_0}{\cos \theta}} . \tag{5.23}$$

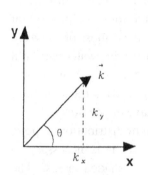

Figure 5.3 Definition of wave number vector.

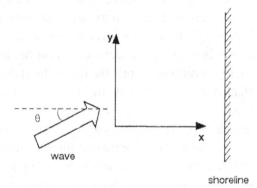

Figure 5.4 Definition of wave propagation.

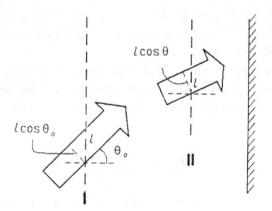

Figure 5.5 Energy flux calculation.

The value $\sqrt{\frac{C_{g0}}{C_g}}$ is the shoaling coefficient K_s in Eq. (5.5) and the refraction coefficient K_r can be defined as

$$K_r = \sqrt{\frac{\cos \theta_0}{\cos \theta}}.$$

(5.24)

5.5 Wave Diffraction

One of the major conditions required in water areas where ports and harbours are constructed, is to create a calm area to protect vessels and infrastructure. Breakwaters are thus constructed to address this need. Figure 5.6 shows how breakwaters affect wave motion. The illustration shows how as the incident waves hit the breakwaters, most of its wave energy is reflected while the rest is dissipated. Due to this water barrier, a shaded area is created, making it ideal for port and harbour zones. However, some of the wave energy in the open area (II) is transmitted to this shaded area (I). This phenomenon is called wave diffraction.

Since it is important to limit the wave height distribution in ports and harbours in order for it to fulfill its purpose, it is necessary to determine the wave height distributions in the shaded areas. The equations used to determine wave height distributions are: Laplace equation (Eq. 3.9), bottom boundary condition (Eq. 3.10), surface boundary conditions (Eq. 3.11, Eq. 3.12) and the condition of no-cross-flow along the breakwater. With these five equations, we can solve the wave height distribution $H(x, y)$ with some appropriate assumptions for the shaded and open areas.

The results of the calculation are obtained by defining a diffraction coefficient K_d, which is the ratio between the absolute value of wave height at specific location $H(x, y)$ and the incident wave height H_0.

$$K_d = \frac{|H(x, y)|}{H_0}.$$

(5.25)

Figure 5.7 shows an example of diffraction coefficient K_d distribution in a case of normal incident wave propagating into a breakwater which has an infinite length in the x-positive direction.

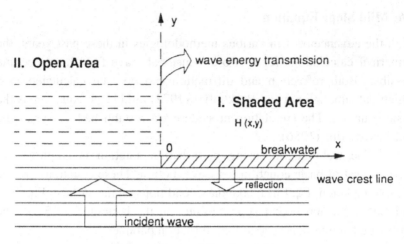

Figure 5.6 Wave diffraction problem.

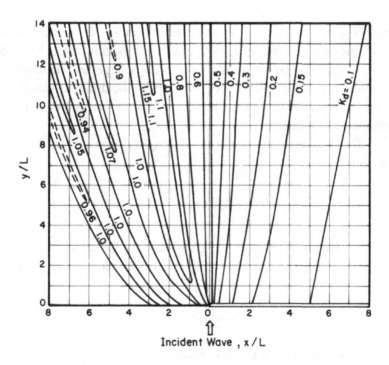

Figure 5.7 Distribution of wave diffraction coefficient K_d (Penney and Price, 1945).

5.6 Mild Slope Equation

With the advancement in various methodologies in these past years, the numerical calculations for a two-dimensional wave field have become possible. Both refraction and diffraction can now be calculated in a numerical simulation using Berkhoff's (1972) mild slope equation as the basic equation. The simulation procedure was established by Watanabe and Maruyayma (1986).

The technique of using perturbation solution is employed to solve the set of basic equation for wave motion. These equations are the Laplace equation, Eq. (3.1), boundary conditions on the surface, Eq. (3.7) and Eq. (3.8), and boundary conditions on the bottom, Eq. (3.5). The following parameter is used as a perturbation parameter.

$$\varepsilon = \frac{H}{\sqrt{\lambda L_1}} \tag{5.26}$$

where, H is wave height, λ is wave length and L_1 is a characteristic length which indicate the reciprocal number of bottom slope. Since the parameter ε should be small, this means L_1 is large and therefore the slope should be small (mild slope). We expand the different variables to series in terms of ε and insert the series into the original equation set (perturbation technique).

If we take up to the first order of $H/\sqrt{\lambda L_1}$, the resultant governing equation is

$$\nabla \cdot (cc_g \nabla \phi) + \sigma^2 \frac{c_g}{c} \phi = 0 \tag{5.27}$$

where c: wave phase velocity, c_g: group velocity, ϕ: velocity potential. If we assume that the surface profile η can be expressed as a periodic function,

$$\eta = ae^{i(\chi - \sigma t)} \tag{5.28}$$

where a: amplitude, χ: phase difference.

If we substitute Eq. (5.28) into Eq. (5.27) and divide the real part and imaginary part, we can get the following simultaneous equations.

From real part:

$$\frac{1}{anc^2}\nabla\cdot(nc^2\nabla a)+k^2-\nabla\chi\cdot\nabla\chi=0. \tag{5.29}$$

From imaginary part:

$$\nabla(nc^2a^2\nabla\chi)=0 \tag{5.30}$$

where $n:c_g/c,\ \nabla\chi=(k\cos\alpha,k\sin\alpha)$.

The numerical solution is to solve these two equations simultaneously.

Figure 5.8 shows the schematic drawing of the numerical result. Here, wave shoaling, refraction, reflection from breakwater and diffraction to the shaded area formed behind breakwater are all included in the numerical calculations.

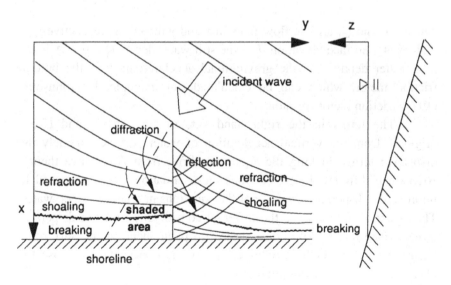

Figure 5.8 Schematic view of wave deformation.

5.7 Wave Field Calculation by Boussinesq Equations (Li and Shibayama, 2000)

The classical forms of the two-dimensional Boussinesq equations are as follows:

$$\frac{\partial \eta}{\partial t} + \frac{\partial Q_x}{\partial x} + \frac{\partial Q_y}{\partial y} = 0 \tag{5.31}$$

$$\frac{\partial Q_x}{\partial t} + \frac{\partial}{\partial x}\left(\frac{Q_x^2}{D}\right) + \frac{\partial}{\partial y}\left(\frac{Q_x Q_y}{D}\right) + gh\frac{\partial \eta}{\partial x} + \frac{f_w}{2h^2}Q_x\sqrt{Q_x^2 + Q_y^2}$$

$$= -\frac{1}{3}h^2\frac{\partial^3 \eta}{\partial x \partial t^2} \tag{5.32}$$

$$\frac{\partial Q_y}{\partial t} + \frac{\partial}{\partial x}\left(\frac{Q_x Q_y}{D}\right) + \frac{\partial}{\partial y}\left(\frac{Q_y^2}{D}\right) + gh\frac{\partial \eta}{\partial y} + \frac{f_w}{2h^2}Q_y\sqrt{Q_x^2 + Q_y^2}$$

$$= -\frac{1}{3}h^2\frac{\partial^3 \eta}{\partial y \partial t^2} \tag{5.33}$$

where, Q_x and Q_y are the flow rates in x and y directions respectively; η is the water surface elevation; h is the still water depth; $D = h + \eta$ is the total water depth; g is the gravitational acceleration; f_w is the bottom friction factor, which can be calculated, for example, by Jonnson's (1974) friction factor formula.

The terms in the right hand side of Eqs. (5.32) and (5.33) originate from the vertical acceleration of water particles, namely the dispersive terms, making the equations differ from Airy wave theory. Afterward, different forms of Boussinesq equations were proposed to improve the dispersive property of the equations in relatively deep water. The essential difference is the different consideration for the dispersive terms. Since this section concentrates on phenomena in the surf zone, where the water depth is relative shallow, it is acceptable to use the classical form of the Boussinesq equations.

To simulate wave decay in the surf zone, it is necessary to introduce momentum dissipation terms in the equations. By introduction of momentum dissipation terms, Eqs. (5.32) and (5.33) become

$$\frac{\partial Q_x}{\partial t} + \frac{\partial}{\partial x}\left(\frac{Q_x^2}{D}\right) + \frac{\partial}{\partial y}\left(\frac{Q_x Q_y}{D}\right) + gh\frac{\partial \eta}{\partial x} + \frac{f_w}{2h^2}Q_x\sqrt{Q_x^2 + Q_y^2}$$

$$= \frac{1}{\rho}\frac{\partial(D\overline{\tau}_{xx})}{\partial x} + \frac{1}{\rho}\frac{\partial(D\overline{\tau}_{xy})}{\partial y} - \frac{1}{3}h^2\frac{\partial^3 \eta}{\partial x \partial t^2} \tag{5.34}$$

$$\frac{\partial Q_y}{\partial t} + \frac{\partial}{\partial x}\left(\frac{Q_x Q_y}{D}\right) + \frac{\partial}{\partial y}\left(\frac{Q_y^2}{D}\right) + gh\frac{\partial \eta}{\partial y} + \frac{f_w}{2h^2}Q_y\sqrt{Q_x^2 + Q_y^2}$$

$$= \frac{1}{\rho}\frac{\partial(D\overline{\tau}_{yx})}{\partial x} + \frac{1}{\rho}\frac{\partial(D\overline{\tau}_{yy})}{\partial y} - \frac{1}{3}h^2\frac{\partial^3 \eta}{\partial x \partial t^2} \tag{5.35}$$

where, ρ is the water density; $\overline{\tau}_{xx}$, $\overline{\tau}_{xy}$, $\overline{\tau}_{yx}$ and $\overline{\tau}_{yy}$ are the vertical-averaged Reynolds stresses, which can be generally expressed as

$$\frac{\overline{\tau}_{ij}}{\rho} = \overline{v}_t\left(\frac{\partial U_i}{\partial x_j} + \frac{\partial U_j}{\partial x_i}\right) - \frac{2}{3}\overline{k}\delta_{ij} \tag{5.36}$$

$$\overline{v}_t = C_\mu \frac{\overline{k}^2}{\overline{\varepsilon}} \tag{5.37}$$

where, $\overline{v}_t$ is the eddy viscosity, $\overline{k}$ is the turbulent energy, $\overline{\varepsilon}$ is the dissipation rate of turbulent energy and the over bar denotes vertical average. The value C_μ is a coefficient, δ_{ij} is the Kronecker operator, U_i and U_j are the vertical-averaged velocity components in i and j direction, and $\overline{k}$ is determined by solving the k equations

$$\frac{\partial \overline{k}}{\partial t} + U\frac{\partial \overline{k}}{\partial x} + V\frac{\partial \overline{k}}{\partial y}$$

$$= \frac{\partial}{\partial x}\left(\frac{\overline{v}_t}{\sigma_k}\frac{\partial \overline{k}}{\partial x}\right) + \frac{\partial}{\partial y}\left(\frac{\overline{v}_t}{\sigma_k}\frac{\partial \overline{k}}{\partial y}\right) + P_h - \overline{\varepsilon} \tag{5.38}$$

where, σ_k is a coefficient with value around 1.0, U and V are the vertical-averaged velocity components in the x and y direction respectively, P_h is the turbulent production supplied by wave breaking, which can be determined by the following formula

$$P_h = \frac{D_V}{\rho d'} \tag{5.39}$$

where, d' is the vertical extent of turbulence, which is approximately equal to the local water depth and D_V is the local dissipation rate of wave energy, which can be calculated by the following formula, proposed on the basis of simple assumption for the ratio of wave height to water depth,

$$D_V = \frac{dE_f}{dx} = \begin{cases} \dfrac{5}{16}\rho g^{\frac{3}{2}} \tan^{\frac{5}{2}} \beta \, \gamma_b^2 x^{\frac{3}{2}}, & x \ge X_m \\[3mm] \dfrac{1}{8}\rho g^{\frac{3}{2}} \tan^{\frac{5}{2}} \beta \, \gamma_1 \gamma_2 x^{\frac{3}{2}}, & x < X_m \end{cases} \tag{5.40}$$

$$\gamma_1 = 2.5\gamma_b + \frac{(5X_m - 13x)(X_m - x)(\gamma_s - \gamma_b)}{2(X_m - X_s)^2} \tag{5.41}$$

$$\gamma_2 = \gamma_b + (\gamma_s - \gamma_b)\left(\frac{X_m - x}{X_m - X_s}\right)^2 \tag{5.42}$$

where, γ_b and γ_s are the ratios of wave height to water depth at breaking point and shoreline, which are 0.8 and 1.5 respectively. The value X_m is the location where the ratio of wave height to water depth deviates from 0.8, which is generally about half of X_b and X_b is the distance from breaking point to shoreline. The value x is the distance from an arbitrary point inside the surf zone to the shoreline and $\tan\beta$ is the bottom slope.

The values of $\bar{\varepsilon}$, $\bar{v}_t$ and $\bar{k}$ are related as follows

$$\bar{\varepsilon} = C_d \bar{k}^{3/2}/l_e \tag{5.43}$$

$$\bar{v}_t = M/C_d^{1/3}\bar{\varepsilon}^{1/3}l_e^{4/3} \tag{5.44}$$

where, l_e is the eddy scale, which is assumed to be equal to local water depth. C_d and M are empirical coefficients.

Turbulence balance is assumed at the shoreline, which means that the dissipated turbulent energy is equal to the loss of wave energy, i.e.

$$\varepsilon_s = P_{hs}. \tag{5.45}$$

The subscript 's' denotes that the value is at the location of the shoreline.

5.8 Examination of Breaker Height Formula (Rattanapitikon, Vivattanasirisak and Shibayama, 2003)

5.8.1 *Introduction*

Because of the complexity of the wave breaking mechanism, it is difficult to describe the breaking wave mechanism by using available wave theories. Hence, the predictions of breaker heights have to be based on empirical formula calibrated from laboratory and field data. However, the validity of empirical formula may be limited according to the range of wave and bottom slope conditions that were employed in the calibrations or verifications. To make an empirical formula reliable, it is necessary to verify the formula with a large amount and wide range of experimental data.

5.8.2 *Existing breaker height formulas*

Most of the existing formulas were developed based on empirical approaches. Some selected breaker height formulas are shown in

Table 5.1 together with their abbreviations. The majority of the existing formulas represent a relationship between the breaking wave height (H_b) and the variables at the deepwater or breaking conditions, i.e., deepwater wavelength (L_o), deepwater wave height (H_o), still water depth at the breaking location (h_b), wavelength at the breaking location (L_b), and bottom slope at the breaking location (m) which can be defined as the change of bottom elevation with respect to distance in cross-shore direction. Here, the term "breaker index" is used to describe non-dimensional breaker height. The existing breaker height formulas shown in Table 5.1 can be classified into three common breaker indices, i.e., the wave height to the depth ratio (H_b/h_b), the wave steepness (H_b/L_b), and the wave height to the deepwater wave height ratio (H_b/H_o). The characteristics of the breaker index derived from Table 5.1 are as follows.

The formulas that are proposed in the form of H_b/h_b are MC94, GA69, CW69, GO70, WE72, MA76, SU80, SW80b, SK88, LK89a, HA90, SK90a, and RS00b. This breaker index seems to be the most frequently re-analyzed topic. The formulas that are proposed in the form of H_b/L_b are MI44, BJ78, OM79, BS85, KA91, and RS00C. The formulas in this group were developed based on the formula of MI44. The formulas that are proposed in the form of H_b/H_o are MK67, KG72, SH74, SW80a, OS84, LK89b, SK90b, GL92, and RS00a.

Rattanapitikon and Shibayama (2000) proposed the following formula by using available laboratory data.

$$H_b = (-1.40m^2 + 0.57m + 0.23)L_b\left(\frac{H_o}{L_o}\right)^{0.35}. \qquad (5.46)$$

A total of 695 cases from 26 sources of published experimental results were collected to develop and verify Eq. (5.46). The experimental data covers a wide range of wave and bottom conditions, ($0.001 \le H_o/L_o \le 0.112$, and $0 \le m \le 0.38$).

Table 5.1 Existing breaker height formulas.

Researchers	Abbreviation	Formulas
McCowan (1894)	MC94	$\dfrac{H_b}{h_b} = 0.78$
Miche (1944)	MI44	$\dfrac{H_b}{L_b} = 0.142 \tanh\left(\dfrac{2\pi h_b}{L_b}\right)$
Le Mehaute and Koh (1967)	MK67	$\dfrac{H_b}{H_o} = 0.76\left(\dfrac{H_o}{L_o}\right)^{-1/4} m^{1/7}$
Galvin (1969)	GA69	$\dfrac{H_b}{h_b} = \dfrac{1}{1.40 - 6.85m}$ for $m \leq 0.07$ $\dfrac{H_b}{h_b} = \dfrac{1}{0.92}$ for $m > 0.07$
Collins and Weir (1969)	CW69	$\dfrac{H_b}{h_b} = (0.72 + 5.6m)$
Goda (1970)	GO70	$\dfrac{H_b}{h_b} = 0.17\dfrac{L_o}{h_b}\left\{1 - \exp\left[-1.5\dfrac{\pi h_b}{L_o}(1 + 15m^{4/3})\right]\right\}$
Weggel (1972)	WE72	$\dfrac{H_b}{h_b} = \dfrac{gT^2 1.56/[1 + \exp(-19.5m)]}{gT^2 + h_b 43.75[1 - \exp(-19m)]}$
Komar and Gaughan (1972)	KG72	$\dfrac{H_b}{H_o} = 0.56\left(\dfrac{H_o}{L_o}\right)^{-1/5}$
Sunamura and Horikawa (1974)	SH74	$\dfrac{H_b}{H_o} = m^{0.2}\left(\dfrac{H_o}{L_o}\right)^{-0.25}$
Madsen (1976)	MA76	$\dfrac{H_b}{h_b} = 0.72(1 + 6.4m)$
Battjes and Janssen (1978)	BJ78	$\dfrac{H_b}{L_b} = 0.14 \tanh\left(\dfrac{0.8}{0.88}\dfrac{2\pi h_b}{L_b}\right)$
Ostendorf and Madsen (1979)	OM79	$\dfrac{H_b}{L_b} = 0.14 \tanh\left[(0.8 + 5m)\dfrac{2\pi h_b}{L_b}\right]$ for $m \leq 0.1$ $\dfrac{H_b}{L_b} = 0.14 \tanh\left[(0.8 + 5(0.1))\dfrac{2\pi h_b}{L_b}\right]$ for $m > 0.1$
Sunamura (1980)	SU80	$\dfrac{H_b}{h_b} = 1.1\left(\dfrac{m}{\sqrt{H_o/L_o}}\right)^{1/6}$
Singamsetti and Wind (1980)	SW80a	$\dfrac{H_b}{H_o} = 0.575 m^{0.031}\left(\dfrac{H_o}{L_o}\right)^{-0.254}$

Table 5.1 (*Continued*)

Researchers	Abbreviation	Formulas
Singamsetti and Wind (1980)	SW80b	$\dfrac{H_b}{h_b} = 0.937 m^{0.155} \left(\dfrac{H_o}{L_o} \right)^{-0.13}$
Ogawa and Shuto (1984)	OS84	$\dfrac{H_b}{H_o} = 0.68 m^{0.09} \left(\dfrac{H_o}{L_o} \right)^{-0.25}$
Battjes and Stive (1985)	BS85	$\dfrac{H_b}{L_b} = 0.14 \tanh \left\{ \left[0.5 + 0.4 \tanh \left(33 \dfrac{H_o}{L_o} \right) \right] \dfrac{2\pi h_b}{0.88 L_b} \right\}$
Seyama and Kimura (1988)	SK88	$\dfrac{H_b}{h_b} = 1.25 \left\{ 0.16 \dfrac{L_o}{h_b} \left\{ 1 - \exp \left[-0.8\pi \dfrac{h_b}{L_o} (1 + 15 m^{4/3}) \right] \right\} - 0.96 m + 0.2 \right\}$
Larson and Kraus (1989)	LK89a	$\dfrac{H_b}{h_b} = 1.14 \left(\dfrac{m}{\sqrt{H_o / L_o}} \right)^{0.21}$
Larson and Kraus (1989)	LK89b	$\dfrac{H_b}{H_o} = 0.53 \left(\dfrac{H_o}{L_o} \right)^{-0.24}$
Hansen (1990)	HA90	$\dfrac{H_b}{h_b} = 1.05 \left(m \dfrac{L_b}{h_b} \right)^{0.2}$
Smith and Kraus (1990)	SK90a	$\dfrac{H_b}{h_b} = \left\{ \dfrac{1.12}{1 + \exp(-60 m)} - 5[1 - \exp(-43 m)] \dfrac{H_o}{L_o} \right\}$
Smith and Kraus (1990)	SK90b	$\dfrac{H_b}{H_o} = (0.34 + 2.47 m) \left(\dfrac{H_o}{L_o} \right)^{-0.30 + 0.88 m}$
Kamphuis (1991)	KA91	$\dfrac{H_b}{L_b} = 0.127 \exp(4 m) \tanh \left(\dfrac{2\pi h_b}{L_b} \right)$
Gourlay (1992)	GL92	$\dfrac{H_b}{H_o} = 0.478 \left(\dfrac{H_o}{L_o} \right)^{-0.28}$
Rattanapitikon and Shibayama (2000)	RS00a	$\dfrac{H_b}{H_o} = (10.02 m^3 - 7.46 m^2 + 1.32 m + 0.55) \left(\dfrac{H_o}{L_o} \right)^{-1/5}$
Rattanapitikon and Shibayama (2000)	RS00b	$\dfrac{H_b}{h_b} = 0.17 \dfrac{L_o}{h_b} \left\{ 1 - \exp \left[\dfrac{\pi h_b}{L_o} (16.21 m^2 - 7.07 m - 1.55) \right] \right\}$
Rattanapitikon and Shibayama (2000)	RS00c	$\dfrac{H_b}{L_b} = 0.14 \tanh \left[(-11.21 m^2 + 5.01 m + 0.91) \dfrac{2\pi h_b}{L_b} \right]$
Rattanapitikon *et al.* (2003)	PS01	$\dfrac{H_b}{L_b} = (-1.40 m^2 + 0.57 m + 0.23) \left(\dfrac{H_o}{L_o} \right)^{0.35}$

5.9 Energy Dissipation of Breaking Waves (Rattanapitikon and Shibayama, 1998)

When waves propagate to the nearshore zone wave profiles steepen and eventually the waves break. Once the waves start to break, a part of the wave energy is transformed to turbulence and heat, and the wave height decreases towards the shore. The rate of energy dissipation of breaking waves is an essential requirement for predicting the wave height, sediment transport rate and beach profile change in the surf zone.

5.9.1 *Energy dissipation for regular waves*

Wave height transformation is computed from the energy flux conservation law. It is given by

$$\frac{\partial(Ec_g \cos \theta)}{\partial x} = -D_B \qquad (5.47)$$

where E is the wave energy density, c_g is the group velocity, θ is the mean wave angle, x is the distance in cross shore direction, x-axis points onshore, and D_B is the energy dissipation rate which is zero outside the surf zone.

Widely used formulas for computing energy dissipation rate are the Bore model and the model of Dally *et al.* (1985). A brief review of these two models is described below.

a) **The Bore model**, originally introduced by Le Mehaute (1962), is developed based on an assumption that the energy dissipation rate of a broken wave is similar to the dissipation rate of a hydraulic jump. Several researchers have proposed slightly different forms of the energy dissipation rate, e.g., Battjes and Janssen (1978):

$$D_B = \frac{\rho g H^2}{4T} = \frac{2}{T} E. \qquad (5.48)$$

Thornton and Guza (1983):

$$D_B = \frac{\rho g H^3}{4Th} = \frac{2H}{Th} E \qquad (5.49)$$

where ρ is the density of water, g is the acceleration due to gravity, H is the wave height, T is the wave period, and h is the water depth.

b) **The model of Dally** *et al.* (1985) is based on the observation of stable wave height on the horizontal bed. They assumed that the energy dissipation rate is proportional to the difference between the local energy flux and the stable energy flux, divided by the local water depth as

$$D_B \propto \frac{[Ec_g - E_s c_g]}{h} \qquad (5.50)$$

or

$$D_B = \frac{K_d c_g}{h}[E - E_s] = \frac{K_d cn}{h}[E - E_s] \qquad (5.51)$$

where

$$E_s = \frac{1}{8}\rho g H_s^2 = \frac{1}{8}\rho g (\Gamma h)^2 \qquad (5.52)$$

$$n = [1 + 2kh/\sinh(2kh)]/2 \qquad (5.53)$$

in which K_d is a constant (decay coefficient), c is the phase velocity, E_s is the stable energy density, H_s is the stable wave height and Γ is the stable wave factor. From the model calibration with the laboratory data of Horikawa and Kuo (1966), Dally *et al.* (1985) found that $K_d = 0.15$ and Γ is varied case by case between 0.35-0.48. However, finally, they suggested to use $\Gamma = 0.4$ for general cases.

Rattanapitikon and Shibayama (1998) assumed the average rate to be proportional to the difference between the local mean energy density and the stable energy density. They obtained the following formula

$$D_B = \frac{0.15c\rho g}{8h}\left[H^2 - \left(h\exp\left(-0.36 - 1.25\frac{h}{\sqrt{LH}}\right)\right)^2\right]. \qquad (5.54)$$

5.9.2 *Energy dissipation for irregular waves*

Irregular wave breaking is more complex than regular wave breaking. In contrast to regular waves there is no well-defined breaking point for irregular waves. The highest waves tend to break at the greatest distances from the shore. Thus, the energy dissipation of irregular waves occurs over a considerably greater area than that of regular waves.

Dally (1992) used the regular wave model of Dally *et al.* (1985) to simulate the transformation of irregular waves by using a wave-by-wave approach. This means that Dally assumed that D_B is proportional to the difference between local energy flux of a breaking wave and stable energy flux. This wave-by-wave approach requires much computation time, and therefore it may not suitable to use in a beach deformation model.

However, the model becomes simple if we set an assumption, similar to that of the present regular wave model, that the average rate of energy dissipation in breaking waves is proportional to the difference between the local mean energy density and stable energy density. After incorporating the fraction of breaking, the average rate of energy dissipation in irregular wave breaking, $\overline{D}_B$, can be expressed as

$$\overline{D}_B = \frac{K_5 Q_b c_p}{h}[E_m - E_s] \tag{5.55}$$

where

$$E_m = \frac{1}{8}\rho g H_{rms}^2 \tag{5.56}$$

$$E_s = \frac{1}{8}\rho g H_s^2 = \frac{1}{8}\rho g (\Gamma_i h)^2 \tag{5.57}$$

in which all variables are computed based on the linear wave theory, where K_5 is the proportional constant, Q_b is the fraction of breaking waves, c_p is the phase velocity related to the peak spectral wave period T_p, h is the water depth, E_m is the local mean energy density, E_s is the stable energy density, H_{rms} is the root mean square wave height, H_s is the stable wave height and Γ_i is the stable wave factor of irregular wave.

Re-writing Eq. (5.55) in term of wave height yields

$$\overline{D}_B = \frac{K_5 Q_b c_p \rho g}{8h} [H_{rms}^2 - (\Gamma_i h)^2].$$ (5.58)

The stable wave factor, Γ_i, is determined as

$$\Gamma_i = \exp\left[K_6 \left(-0.36 - 1.25 \frac{h}{\sqrt{L_p H_{rms}}} \right) \right]$$ (5.59)

where K_6 is the coefficient, L_p is the wavelength related to the peak spectral wave period. Then Eq. (5.58) becomes

$$\overline{D}_B = \frac{K_5 Q_b c_p \rho g}{8h} \left[H_{rms}^2 - \left(h \exp\left(-0.36 K_6 - 1.25 K_6 \frac{h}{\sqrt{L_p H_{rms}}} \right) \right)^2 \right].$$ (5.60)

The local fraction of breaking waves, Q_b, is determined from the derivation of Battjes and Janssen (1978) based on the assumption of a truncated Rayleigh distribution at the maximum wave height

$$\frac{1 - Q_b}{-lnQ_b} = \left(\frac{H_{rms}}{H_b} \right)^2$$ (5.61)

where H_b is the breaking wave height that can be computed by using the breaking criteria of Goda (1970)

$$H_b = K_7 L_o \left\{ 1 - \exp\left[-1.5 \frac{\pi h}{L_o} (1 + 15 m^{4/3}) \right] \right\}$$ (5.62)

where K_7 is a constant, L_o is the deep-water wavelength related to the peak spectral wave period, and m is the bottom slope.

Since Eq. (5.61) is an implicit equation, an iteration process is necessary to compute the fraction of breaking waves, Q_b. It would be more convenient if we could compute Q_b from the explicit form of

Eq. (5.61). From the multi-regression analysis, the explicit form of Q_b can be expressed as the following (with $R^2 = 0.999$)

$$Q_b = \begin{cases} 0 & for \ \dfrac{H_{rms}}{H_b} \le 0.43 \\[2ex] -0.738\left(\dfrac{H_{rms}}{H_b}\right) - 0.280\left(\dfrac{H_{rms}}{H_b}\right)^2 + 1.785\left(\dfrac{H_{rms}}{H_b}\right)^3 + 0.235 \\[2ex] & for \ \dfrac{H_{rms}}{H_b} > 0.43 \end{cases} \qquad (5.63)$$

Trial simulations indicated that $K_5 = 0.10$, $K_6 = 1.60$, and $K_7 = 0.10$ give good agreement between measured and computed *rms* wave heights. Finally, the energy dissipation rate of irregular wave breaking can be written as

$$\bar{D}_B = \frac{0.1 Q_b c_p \rho g}{8h}\left[H_{rms}^2 - \left(h \exp\left(-0.58 - 2.00\frac{h}{\sqrt{L_p H_{rms}}} \right) \right)^2 \right]. \qquad (5.64)$$

References

Battjes, J.A. (1974): Surf similarity, *Proc. 14th Coastal Eng. Conf.*, pp. 466-480.

Battjes, J.A. and Janssen, J.P.F.M. (1978): Energy loss and set-up due to breaking of random waves, *Proc. 16th Coastal Eng. Conf.*, ASCE, pp. 569-589.

Battjes, J.A. and Stive, M.J.F. (1985): Calibration and verification of a dissipation model for random breaking waves, *J. Geophysical Research 90*, C5, pp. 9159-9167.

Berkhoff, J.C.W. (1972): Computation of combined refraction-diffraction, *Proc. 13th Coastal Eng. Conf.*, ASCE, pp. 471-490.

Collins, J.I. and Weir, W. (1969): Probabilities of wave characteristics in the surf zone, *Tetra Tech. Report*, TC-149, Pasadena, California, USA, 122 p.

Dally, W.R., Dean, R.G. and Dalrymple, R.A. (1985): Wave height variation across beach of arbitrary profile, *J. of Geo. Res.*, Vol. 90, No. C6, pp. 11917-11927.

Dally, W.R. (1992): Random breaking waves: Field verification of a wave-by-wave algorithm for engineering application, *Coastal Eng.*, Vol. 16, pp. 369-397.

Galvin, C.J. (1969): Breaker travel and choice of design wave height, *J. Waterways and Harbors Division*, ASCE 95, WW2, pp. 175-200.

Goda, Y. (1970): A synthesis of breaker indices, *Trans. Japan Society of Civil Engineers 2*, pp. 227-230.

Goda, Y. (1975): Deformation of irregular waves due to depth-controlled wave breaking, *Report of the Port and Harbour Research Institute*, Vol. 14, No. 3.

Gourlay, M.R. (1992): Wave set-up, wave run-up and beach water table: Interaction between surf zone hydraulics and groundwater hydraulics, *Coastal Eng. 17*, pp. 93-144.

Horikawa, K. and Kuo, C.T. (1966): A study of wave transformation inside surf zone, *Proc. 10th Coastal Engineering Conf.*, ASCE, pp. 217-233.

Hunt, I.A., Jr. (1961): Design of seawalls and breakwaters, *Trans. of ASCE*, 126, IV, pp. 542-570.

Kamphuis, J.W. (1991): Incipient wave breaking, *Coastal Eng. 15*, pp. 185-203.

Karambas, T.V. and Koutitas, C. (1992): A breaking wave propagation model based on the Boussinesq equations, *Coastal Eng. 18*, pp. 1-19.

Komar, P.D. and Gaughan, M.K. (1972): Airy wave theory and breaker height prediction, *Proc. 13rd Coastal Eng. Conf.*, ASCE, pp. 405-418.

Larson, M. and Kraus, N.C. (1989): SBEACH: Numerical model for simulating storm-induced beach change, Report 1, *Tech. Report CERC-89-9*, Waterways Experiment Station, US Army Corps of Engineers, 267 p.

Le Mehaute, B. and Koh, R.C.Y. (1967): On the breaking of waves arriving at an angle to the shore, *J. Hydraulic Research 5*, 1, pp. 67-88.

Li Shaowu, Wang Shang-yi and Shibayama Tomoya (1998): A nearshore wave breaking model, *ACTA Oceanology Sinica 17*, No. 1, pp. 108-118.

Madsen, O.S. (1976): *Wave climate of the continental margin: Elements of its mathematical description*, D.J. Stanley and D.J.P. Swift (Editors), Marine Sediment Transport in Environmental Management. Wiley, New York, pp. 65-87.

McCowan, J. (1894): On the highest waves of a permanent type, *Philosophical Magazine*, Edinburgh 38, 5th Series, pp. 351-358.

Miche, R. (1944): Mouvements ondulatoires des mers en profondeur constante on decroissante, *Ann. des Ponts et Chaussees*, Ch. 114, pp. 131-164, 270-292, and 369-406.

Mizuguchi, M. (1980): An heuristic model of wave height distribution in surf zone, *Proc. 17th Coastal Eng. Conf.*, pp. 278-289.

Ogawa, Y. and Shuto, N. (1984): Run-up of periodic waves on beaches of non-uniform slope, *Proc. 19th Coastal Eng. Conf.*, ASCE, pp. 328-344.

Ostendorf, D.W. and Madsen, O.S. (1979): *An analysis of longshore current and associated sediment transport in the surf zone*, Report No. 241, Dept. of Civil Eng., MIT, 169 p.

Rattanapiticon, W. and Shibayama, T. (1998): Energy dissipation model for regular and irregular breaking waves, *Coastal Eng. Journal*, 40(4), 327-346.

Rattanapitikon, W. and Shibayama, T. (2000): Verification and modification of breaker height formulas, *Coastal Eng. Journal*, JSCE, 42(4), pp. 389-406.

Rattanapitikon, W., Vivattanasirisak, T. and Shibayama, T. (2003): Proposal of new breaker height formula, *Coastal Eng. Journal*, 45(1), 29-48.

Sato, S. and Kabiling, M. (1993): Numerical calculation of three-dimensional bed deformation on basis of Boussinesq equations, *Proc. Conf. on Coastal Eng.*, JSCE, Vol. 40, pp. 386-390 (in Japanese).

Seyama, A. and Kimura, A. (1988): The measured properties of irregular wave breaking and wave height change after breaking on slope, *Proc. 21st Coastal Eng. Conf.*, ASCE, pp. 419-432.

Singamsetti, S.R. and Wind, H.G. (1980): *Characteristics of breaking and shoaling periodic waves normally incident on to plane beaches of constant slope*, Report M1371, Delft Hydraulic Lab., Delft, The Netherlands, 142 p.

Smith, J.M. and Kraus, N.C. (1990): *Laboratory study on macro-features of wave breaking over bars and artificial reefs*, Technical Report CERC-90-12, WES, U.S. Army Corps of Engineers, 232 p.

Sunamura, T. and Horikawa, K. (1974): Two-dimensional beach transformation due to waves, *Proc. 14th Coastal Eng. Conf.*, ASCE, pp. 920-938.

Sunamura, T. (1980): A laboratory study of offshore transport of sediment and a model for eroding beaches, *Proc. 17th Coastal Eng. Conf.*, ASCE, pp. 1051-1070.

Thornton, E.B. and Guza, R.T. (1983): Transformation of wave height distribution, *J. Geophys. Res.*, Vol. 88, pp. 5925-5983.

Watanabe, A. and Maruyama, K. (1986): Numerical modeling of nearshore wave field under combined refraction, diffraction and breaking, *Coastal Eng. in Japan*, Vol. 29, pp. 19-30.

Weggel, J.R. (1972): Maximum breaker height, *J. Waterways, Harbors and Coastal Eng. Div. 98*, WW4, pp. 529-548.

Exercise

Problem 5.1

1) Show the governing equation and boundary conditions for wave motion.

2) Derive linearized equations and boundary conditions and give reasons for any assumptions that you make.

3) Then derive expressions for particle velocity (horizontal and vertical direction) and pressure field. (Small amplitude wave theory)

Problem 5.2

Explain the following items briefly by using formulas, figures and descriptions.

1) Snell's law for wave refraction.
2) The difference between spilling breaker and plunging breaker.
3) The importance of surf zone.

Problem 5.3

A wave generator is installed at one end of a wave flume and is generating waves. After some time, a quasi-steady condition has been reached and from one wave gage placed at 25 cm intervals along the wave flume, it is observed that the wave height varies with distance x from the generator, in the following manner.

x (cm)	200	225	250	275	300	325	350	375	400
H (cm)	5.3	6.0	5.3	3.5	2.0	3.5	5.3	6.0	5.3

At the other end of the flume, a coastal structure was installed.

1) What is the reflection coefficient of the structure? (Healey's method)
2) What is the incident wave height?
3) What is the reflection coefficient at the structure?
4) With the water depth being 30 cm, what is the period T of the wave maker motion?

(Modified from the homework problem of Prof. Madsen in 1977.)

Problem 5.4

Waves normally incident on a straight gently sloping beach are recorded to have a wave height of 1 m and a period of 8 s in water depth of 3 m. Assume that linear theory is valid and neglect reflection and energy dissipation.

1) What is the wave height and wave length in the deep water condition?

2) Use linear wave theory and the simplified breaking criteria $(H_b/h_b) = 0.8$, and predict the water depth and wave height at breaking point (calculate wave shoaling from offshore to onshore direction and compare with the breaking criteria).

Waves with wave height of 0.88 m. and a period of 8 s. are obliquely incident on a long straight beach of gentle slope. The angle of incident in deep water is 45 degrees.

3) Use linear wave theory and estimate the wave height and angle of incidence in water depth of 8 m and 3 m. Compare the results with the results of 1) and explain the reason for the differences.

(Modified from the homework problem of Prof. Madsen in 1977.)

Problem 5.5

A current meter was placed at the bottom of water where the still water level is 20 m. The meter records two horizontal velocity components u and v in the x and y direction, respectively. A simple unidirectional (longcrested swell) wave passes over the current meter which records the following velocity time history;

$u = u_0 \cos \sigma t$

$v = v_0 \cos \sigma t$ (These two records are in phase.)

From the records it is seen that the wave period is 6 s and $u_0 = 0.1 m/s$ and $v_0 = 0.2 m/s$.

1) Calculate the direction of propagation of the wave (angle with x-axis of current meter).
2) Calculate the wave height and wave length.

(Modified from the homework problem of Prof. Madsen in 1977.)

Chapter 6

Surfzone Dynamics

The area between the wave breaking line and the shoreline is called the surfzone. In this area, wave energy transported from the offshore is transformed into turbulent energy and then into heat energy and is therefore dissipated. Analysis of this energy dissipation process started in the middle of 1970s and the recent research has shown that wave height distribution can be simulated by numerical models.

6.1 General View of Surfzone Dynamics

There are basically two possible causes for energy dissipation in the surfzone: bottom friction dissipation and turbulent energy dissipation due to breaking waves. The bottom friction effect was already discussed in Section 4.2, and therefore, in this section, the energy dissipation model under breaking waves will be explained.

Mizuguchi (1980) modeled the energy dissipation rate ε as a function of the eddy viscosity coefficient v_e, wave number k and wave height H. In his model,

$$\varepsilon = -\frac{d}{dx}(EC_g) \qquad (6.1)$$

where $E = \frac{1}{8}\rho g H^2$ and C_g is the wave group velocity. And

$$\varepsilon = 2\rho g v_e \left(k\frac{H}{2}\right)^2. \qquad (6.2)$$

The value v_e is taken as an empirical constant. By using Eq. (6.1) and (6.2), the wave height change can be well simulated for the conditions of a uniform slope beach, step type beach or bar type beach.

One more important problem is wave run-up. At the shoreline boundary between land and water, the run-up wave from the water area comes and moves over the dry bed slope, stops and returns to the water area. It is an important engineering problem to estimate the maximum height of water run-up. For the topography shown in Fig. 6.1, Hunt (1961) presented an empirical formula to calculate the maximum wave run-up. Run-up height R can be estimated as

$$\frac{R}{H} = \frac{1.01 \tan \theta}{(H/L_0)^{\frac{1}{2}}}.$$
(6.3)

R: run-up height, H: wave height in constant depth, L_0: deep water wave length and $\tan \theta$: bottom slope.

Equation (6.3) was originally derived for the wave run-up over the slope of an artificial breakwater to estimate the rate of wave over-topping. It can be applied to natural beaches if the beach profile can be modeled as per Fig. 6.1.

It is also important to estimate the velocity field in the surfzone for sediment transport, diffusion and dispersion processes, or other environmental related problems. In general, the velocity time history for the surfzone can be divided into four components: steady flow, U, wave component u_w, large vortex component u_v and turbulence u'. Among these four, the wave and large vortex components are periodic functions

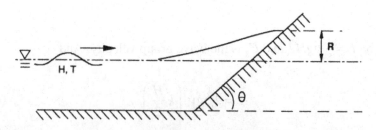

Figure 6.1 Wave run-up.

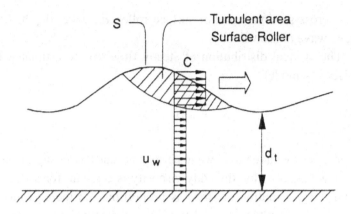

Figure 6.2 Propagation of surface roller.

of time which have the same period as the incident wave to the surfzone. The large vortex component comes from the systematic vortex motion formed under breaking waves.

The way to estimate the steady flow field is described as follows (Okayasu *et al.*, 1988). For an on-offshore distribution of depth-averaged steady flow, there are two contributions. The first one is from the wave motion and the resultant mass transport due to wave action. The second one is from surface roller propagation. The surface roller is a vortex which travels with the breaking wave crest and produces high turbulence which characterizes the turbulent nature of the surfzone (see Fig. 6.2). With the propagation of the surface roller, the water mass is transported and this results in a steady flow. From the mass flux conservation, the total flux of offshore-directed steady flow U_{off} is equal to the total flux onshore due to wave U_w and surface roller U_S.

$$U_{off} = -U_w - U_S \qquad (6.4)$$

where U_{off}: offshore steady flow, U_w: onshore steady flow derived from wave theory, U_S: onshore steady flow due to surface roller.

The value U_S can be expressed as

$$U_S = \frac{S}{d_t L} c \qquad (6.5)$$

where S: cross-sectional area of surface roller, d_t: water depth, L: wave length, c: wave phase velocity.

The vertical distribution of steady flow can be estimated by the eddy viscosity model.

$$\bar{\tau} = \rho v_t \frac{\partial U}{\partial z} \tag{6.6}$$

$\bar{\tau}$: shear stress averaged over wave period, v_t: eddy viscosity in surfzone.

If we assume that the eddy viscosity is constant for a fixed value of X and the distribution of stress is proportional to the elevation from the bottom, the resultant vertical profile of a steady flow is a parabolic distribution. The offshore directed steady flow in the bottom area is called undertow.

6.2 Numerical Model of Surfzone Velocity Field (Shibayama and Duy, 1994)

A numerical model for both non-breaking and breaking waves was proposed by Shibayama and Duy (1994), in which the governing equations are based on equi-phase averaged Navier-Stokes equations, or the Reynolds equations. Transforming a variable physical domain to a fixed computational domain solves the model. Verification of the model with various sets of laboratory data shows that, through this model, it is possible to simulate fairly reasonably the nonlinear and asymmetric characteristics of the wave parameters in a vertical plane on a sloping bottom in terms of the wave shape and water particle velocities. The deformation of the wave profile due to the shoaling effect, the vertical distribution of pressure and water particle velocities can be studied in detail. The steady current induced by the breaking waves inside the surf zone can also be computed by using a vertical distribution of the water particle velocities.

6.2.1 *Hydrodynamic model*

The breaker-generated turbulent flow in the surf zone is modeled by using the two-dimensional vertical model, in which the equi-phase mean quantities of turbulent motion (water surface, pressure, velocities) are determined simultaneously by the solution of the equi-phase-averaged Navier-Stokes equations, also referred to as the Reynolds equations of motion.

$$\frac{\partial u}{\partial x} + \frac{\partial w}{\partial z} = 0 \tag{6.7}$$

$$\frac{\partial u}{\partial t} + \frac{\partial (u^2)}{\partial x} + \frac{\partial (uw)}{\partial z} = -\frac{1}{\rho}\frac{\partial P}{\partial x} + v\left(\frac{\partial^2 u}{\partial x^2} + \frac{\partial^2 u}{\partial z^2}\right) + M_x \tag{6.8}$$

$$\frac{\partial w}{\partial t} + \frac{\partial (uw)}{\partial x} + \frac{\partial (w^2)}{\partial z} = -g - \frac{1}{\rho}\frac{\partial P}{\partial z} + v\left(\frac{\partial^2 w}{\partial x^2} + \frac{\partial^2 w}{\partial z^2}\right) + M_z \tag{6.9}$$

$$\frac{\partial \zeta}{\partial t} + \frac{\partial}{\partial x}\int_{z_b}^{\zeta} u\,dz = 0 \tag{6.10}$$

where u, w: equi-phase mean velocities in x and z-directions; P: equi-phase mean total pressure; ζ: z-coordinate of water surface; z_b: z-coordinate of sea bottom; g: gravity acceleration; v: molecular viscosity; M_x, M_z: momentum transports due to turbulence in x and z-directions (Reynolds stress terms).

$$M_x = 2\frac{\partial}{\partial x}\left(v_t \frac{\partial u}{\partial x}\right) + \frac{\partial}{\partial z}\left[v_t\left(\frac{\partial u}{\partial z} + \frac{\partial w}{\partial x}\right)\right] \tag{6.11}$$

$$M_z = \frac{\partial}{\partial x}\left[v_t\left(\frac{\partial u}{\partial z} + \frac{\partial w}{\partial x}\right)\right] + 2\frac{\partial}{\partial z}\left(v_t \frac{\partial w}{\partial z}\right) \tag{6.12}$$

where v_t is the time-dependent eddy viscosity, which is derived based on the distribution of the mean eddy viscosity proposed by Svendsen and Hansen (1988) for undertow modelling.

$$v_t = f_v\sqrt{gh}(\zeta - z_b). \tag{6.13}$$

In Eq. (6.13), f_v is a constant and was found to have an average value of 0.125 from the computations of breaking waves (Shibayama and Duy, 1994). In order to take into account the fact that the eddy viscosity commonly increases within certain distance from the breaking point, as shown by the laboratory measurements of Okayasu *et al.* (1988), the constant f_v is slightly modified in the present section as follows.

$$f_v = \begin{cases} 0.03e^{h_b/h} & \text{if } h_b/h < 1.5 \\ 0.03e^{1.5} = 0.134 & \text{if } h_b/h \geq 1.5 \end{cases} \quad (6.14)$$

where h is the still water depth, and h_b the breaking water depth. Eq. (6.14) expresses a variation range of f_v from a minimum value of 0.082 at the breaking point ($h_b/h = 1$) to a maximum value of 0.134. It has been verified from various computations that the above expression of f_v gives good simulation results of the velocity field in the surf zone, in particular for the near-breaking area.

With the use of general and highly nonlinear governing equations, the above hydrodynamic model is capable of simulating the nonlinear and asymmetric characteristics of time-dependent wave parameters after wave breaking, as shown by Shibayama and Duy (1994).

6.2.2 *Numerical formulation*

The equation is solved by a similar method to that used to solve the breaking-wave model (Shibayama and Duy, 1994). This solution method is based on a transformation technique which maps the moving and nonlinear physical domain (x, y, t) with a fixed and linear computational domain (ξ, η, τ) (see Fig. 6.3). The functional relationships between the two domains is given by

$$\xi = x \quad (6.15)$$

$$\eta = H_s \frac{z - z_b(x)}{\zeta(x,t) - z_b(x)} \quad (6.16)$$

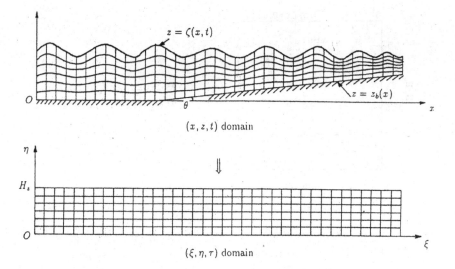

Figure 6.3 Physical and computational domains.

$$\tau = t \tag{6.17}$$

where H_s is the maximum vertical length in the computational domain.

The first derivatives of the velocity components, u and w, with respect to x, z and t are expressed in terms of the new variables ξ, η and τ as follows.

$$\begin{bmatrix} \dfrac{\partial u}{\partial x} & \dfrac{\partial u}{\partial z} & \dfrac{\partial u}{\partial t} \\[2ex] \dfrac{\partial w}{\partial x} & \dfrac{\partial w}{\partial z} & \dfrac{\partial w}{\partial t} \end{bmatrix} = \begin{bmatrix} \dfrac{\partial u}{\partial \xi} & \dfrac{\partial u}{\partial \eta} & \dfrac{\partial u}{\partial \tau} \\[2ex] \dfrac{\partial w}{\partial \xi} & \dfrac{\partial w}{\partial \eta} & \dfrac{\partial w}{\partial \tau} \end{bmatrix} \begin{bmatrix} \dfrac{\partial \xi}{\partial x} & \dfrac{\partial \xi}{\partial z} & \dfrac{\partial \xi}{\partial t} \\[2ex] \dfrac{\partial \eta}{\partial x} & \dfrac{\partial \eta}{\partial z} & \dfrac{\partial \eta}{\partial t} \\[2ex] \dfrac{\partial \tau}{\partial x} & \dfrac{\partial \tau}{\partial z} & \dfrac{\partial \tau}{\partial t} \end{bmatrix} \tag{6.18}$$

where the Jacobian matrix $\mathbf{J}$ of the transformation is determined by using Eqs. (6.15) through (6.17) as follows.

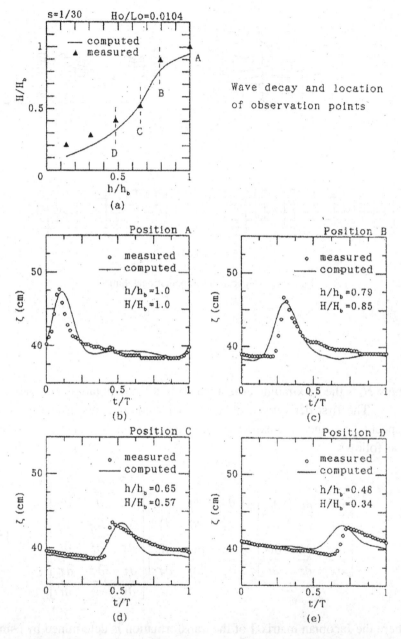

Figure 6.4 Comparison of wave profile inside surf zone (measurements by Okayasu *et al.*, 1988: $H_0 = 6.30$ cm, $T = 1.97s$, $h_0 = 40$ cm, $s = 1/30$).

$$
\mathbf{J} = \begin{bmatrix} \dfrac{\partial \xi}{\partial x} & \dfrac{\partial \xi}{\partial z} & \dfrac{\partial \xi}{\partial t} \\[2mm] \dfrac{\partial \eta}{\partial x} & \dfrac{\partial \eta}{\partial z} & \dfrac{\partial \eta}{\partial t} \\[2mm] \dfrac{\partial \tau}{\partial x} & \dfrac{\partial \tau}{\partial z} & \dfrac{\partial \tau}{\partial t} \end{bmatrix}
$$

$$
= \begin{bmatrix} 1 & 0 & 0 \\[2mm] -\dfrac{[H_s z_{bx} + \eta(\zeta_x - z_{bx})]}{\zeta - z_b} & \dfrac{H_s}{\zeta - z_b} & -\eta \dfrac{\zeta_t}{\zeta - z_b} \\[2mm] 0 & 0 & 1 \end{bmatrix}. \qquad (6.19)
$$

Figures 6.4(b) through 6.4(e) show the time series of the surface elevation at selected positions within the surf zone as indicated in Fig. 6.4(a) and the comparison with the measurements by Okayasu *et al.* (1988). The measured wave profiles exhibit steeper crests than the computed ones, and the discrepancy between the two wave profiles becomes larger as the wave moves closer to the shoreline.

6.3 A Three-Dimensional Large Eddy Simulation - LES (the following description are based on Suzuki, Okayasu and Shibayama, 2006)

A three-dimensional Large Eddy Simulation (LES) model was proposed by Suzuki *et al.* (2006) together with a sediment pickup/advection module for the study of spatial and temporal variation of sediment concentration due to three-dimensional fluid motion under breaking waves on a sloping beach. The LES flow module calculates the turbulent flow due to wave breaking where a large-scale turbulence can be directly solved by the governing flow equations and a small-scale turbulence can be evaluated using a sub-grid-scale (SGS) model.

6.3.1 *Methodology*

Several numerical models can simulate the velocity field in the surf zone. For an accurate calculation of this velocity field, the effect of complex fluid motion and turbulence dissipation need to be included. However, only a few models can represent the velocity field of complex fluid motion and the effects of high turbulence dissipation, such as wave breaking and organized eddies.

The Direct Numerical Simulation (DNS) model can be used to calculate fluid motions including small scales of turbulence (e.g. Wijayaratna and Okayasu, 2000, Brasseur and Lin, 2005). However, because it requires a fine grid to resolve the turbulence, its application is limited. In contrast, the effect of turbulence can be modeled using the Reynolds-Averaged Numerical Simulation (RANS) model. Lin and Liu (1998) coupled the RANS equation with a second-order k-ε turbulence model to describe spilling and plunging breaking waves. Chang *et al.* (2001, 2005) also used a two-dimensional RANS equation with a k-ε closure to study the vortex generation due to flow separation around a submerged reactor obstacle.

The Large Eddy Simulation (LES) model combines these two schemes. The turbulence is separated by its scale length depending upon whether it is larger or smaller than the computational grid size. A larger scale of turbulence is solved directly using Navier-Stokes equations while a smaller scale turbulence can be solved using a turbulence equation. Therefore, the LES model can potentially resolve velocity fluid in the surf zone including all scales of turbulence and can also reduce the computational domain and time length in comparison with DNS models.

The flow field is solved by a one-phase (liquid-phase) three-dimensional Large Eddy Simulation (LES). The governing equations are the spatially filtered three-dimensional Navier-Stokes equations along with the continuity equation.

The total flow velocity, u_i (i = 1, 2, 3) can be decomposed into two parts in accordance with the representative scale length of the turbulence. A large-scale turbulence, whose representative length is larger than the grid size, such as large organized eddies due to wave breaking, is directly evaluated by solving the spatially averaged Navier-Stokes equation and the continuity equation for an incompressible fluid:

$$\frac{\partial \bar{u}_i}{\partial t} + \bar{u}_j \frac{\partial \bar{u}_i}{\partial x_j} = -\frac{1}{\rho}\frac{\partial \bar{p}}{\partial x_i} + \nu \frac{\partial^2 \bar{u}_i}{\partial x_j \partial x_j} - \frac{\partial}{\partial x_j}(\tau_{ij}) + g_i \qquad (6.20)$$

$$\frac{\partial \bar{u}_i}{\partial x_i} = 0 \qquad (6.21)$$

where $\bar{u}_i$ is the velocity component spatially averaged over the grid size, t is time, x_i ($i = 1, 2, 3$) is the coordinate spacing, ρ is the density of fluid, p is the pressure, v is the molecular viscosity, g_i is the i th component of the gravitational acceleration, and τ_{ij} is the sub-grid-scale (SGS) stress.

The velocity component of the small-scale turbulence, the representative length of which is smaller than the grid size, is accounted for in the equation as the SGS stress:

$$\tau_{ij} = \rho \overline{u_i' u_j'} = \rho \, (\overline{u_i u_j} - \bar{u}_i \bar{u}_j) \qquad (6.22)$$

where u_i' is the small scale turbulence of the velocity component.

$$u_i = \bar{u}_i + u_i'. \qquad (6.23)$$

In order to estimate the SGS stress τ_{ij}, several models such as the Smagorinsky Model (Smagorinsky, 1963), Dynamic Smagorinsky Model (Germano *et al.*, 1991) and Structure Function Model (Lesieur and Metais, 1996) have been presented. The method outlined in the present book uses the Smagorinsky model, which has been widely used in many LES models (e.g. Christensen and Deigaard, 2001, Okayasu *et al.*, 2005). With this model, the SGS stress is expressed as:

$$\tau_{ij} = -2v_e \bar{D}_{ij} \qquad (6.24)$$

where v_e is the viscosity coefficient for the SGS model and $\bar{D}_{ij}$ is the strain rate. These are given by:

$$v_e = (C_s \Delta)^2 \sqrt{2 \bar{D}_{ij} \bar{D}_{ij}} \qquad (6.25)$$

$$\bar{D}_{ij} = \frac{1}{2} \left(\frac{\partial \bar{u}_i}{\partial x_j} + \frac{\partial \bar{u}_j}{\partial x_i} \right). \qquad (6.26)$$

Here C_s is the Smagorinsky constant and Δ is the spatial length scale:

$$\Delta = (\Delta x_1 \Delta x_2 \Delta x_3)^{1/3} \qquad (6.27)$$

where Δx_i is the grid spacing in the i th direction, which is 1 cm. Schumann (1987) suggested that the value of the Smagorinsky constant C_s is between 0.07 and 0.21. The value can be empirically determined by calibration using experimental data. However, a value of 0.1 has been used in various studies (e.g. Christensen and Deigaard, 2001) and is also adopted here.

The Cubic Interpolated Pseudo-particle (CIP) method (Yabe *et al.*, 1990) is employed to solve the finite difference scheme of the governing equations, and the Successive Over-Relaxation (SOR) method (Hino *et al.*, 1983, 1985) is used to solve the pressure equation. The free surface position is calculated using the density function method based on the work of Watanabe and Saeki (1999). The density function f is determined for every grid cell and expresses the ratio of water ($\rho = 1$ g/cm^3) to the volume of the cell. If a grid cell is filled with water, the function takes 1, but is 0 with no water in the cell. Thus, $f \geq 0.5$ indicates a water cell and $f < 0.5$ indicates an air cell. The water surface is assumed to be located on a curve where the function equals 0.5. In the calculation, the density function f is renewed at every time step with the equation below.

$$\frac{Df}{Dt} = 0. \qquad (6.28)$$

The irregular star method (Chan and Street, 1970) is adopted to evaluate the pressure at each grid cell where the water surface is located.

6.3.2 *Numerical results*

Figure 6.5 shows successive contour maps of the sediment concentration superimposed on the flow velocity vectors calculated in a cross-shore vertical section at the middle of the flume ($y = 10$ cm). To facilitate visualization, the cross-shore distance between vectors is set at 8 cm. The

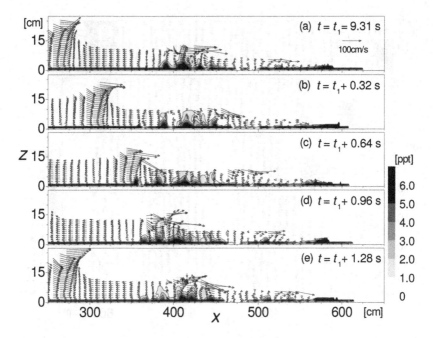

Figure 6.5 Snapshot of velocity vectors and sediment concentration distribution in a cross-shore vertical section at the middle of the flume ($y = 10$ cm) (Suzuki *et al.*, 2006).

complex flow field generated by wave breaking appears to be well represented by the model in the qualitative sense, and the special and temporal intermittency of the sediment suspension and/or re-suspension can be simulated. Also, around the break impact point ($x = 360\text{-}480$ cm), high concentrated sediment clouds are intermittently transported from the lower layer to the water surface by the fluid motion.

Figure 6.6 presents snapshots of the sediment pickup concentration pattern from the bottom boundary to the lowest cells in the horizontal section (tilted $s = 1{:}20$) together with velocity vectors in this section. To facilitate visualization, the cross-shore distance between vectors is set at 8 cm. From the figure, three different patterns for sediment pickup can be seen, namely at the breaking point, at the break impact area, and at the run-up zone.

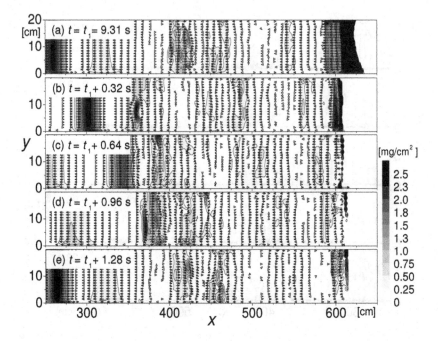

Figure 6.6 Snapshot of velocity vectors and sediment pickup concentration in the bottom horizontal plane (tilted 1:20) (Suzuki *et al.*, 2006).

References

Anderson, D.A., Tannehill, J.C. and Pletcher, R.H. (1984): *Computational fluid mechanics and heat transfer*, Hemisphere Publishing Corp., New York, pp. 197-205.

Battjes, J.A. (1974): Surf similarity, *Proc. 14th Coastal Eng. Conf.*, pp. 466-480.

Berkhoff, J.C.W. (1972): Computation of combined refraction-diffraction, *Proc. 13th Coastal Eng. Conf.*, ASCE, pp. 471-490.

Brasseur, J.G. and Lin, W. (2005): Kinematics and dynamics of small-scale vorticity and strain-rate structures in the transition from isotropic to shear turbulence, *Fluid Dynamics Res.*, 36, pp. 357-384.

Chan, R.K.-C. and Street, R.L. (1970): A computer study of finite-amplitude water waves, *J. Comp. Phys.*, 6, pp. 68-94.

Chang, K.-A., Hsu, T.-J. and Liu, P.L.-F. (2001): Vortex generation and evolution in water waves propagating over a submerged rectangular obstacle: Part I: Solitary waves, *Coastal Eng.*, 44, pp. 13-36.

Chang, K.-A., Hsu, T.-J. and Liu, P.L.-F. (2005): Vortex generation and evolution in water waves propagating over a submerged rectangular obstacle: Part II: Cnoidal waves, *Coastal Eng.*, 52, pp. 257-283.

Christensen, E.D. and Deigaard, R. (2001): Large eddy simulation of breaking waves, *Coastal Eng.*, 42, pp. 53-86.

Cox, D.T. and Kobayashi, N. (2000): Identification of intense, intermittent coherent motions under shoaling and breaking waves, *J. Geophys. Res.*, Vol. 105, No. C6, pp. 14223-14236.

Fletcher, C.A.J. (1991): *Computational techniques for fluid dynamics*, Vol. 1, 2, 2nd edition, Springer-Verlag, Berlin Heidelberg, pp. 299-328.

Germano, M., Piomelli, U., Moin, P. and Cabot, W.H. (1991): A dynamic subgrid-scale eddy viscosity model, *Phys. Fluids*, A, Vol. 3, No. 7, pp. 1760-1765.

Goda, Y. (1975): *Deformation of irregular waves due to depth-controlled wave breaking*, Report of the Port and Harbour Research Institute, Vol. 14, No. 3.

Hino, T., Miyata, H. and Kajitani, H. (1983): A numerical solution method for nonlinear shallow water waves, *J. of Soc. Naval Arch. Japan*, Vol. 153, pp. 1-12.

Hino, T., Miyata, H. and Kajitani, H. (1985): Numerical simulation of nonlinear behavior of three-dimensional ocean waves interacting with obstacle, *J. of Soc. Naval Arch. Japan*, Vol. 157, pp. 141-154.

Hunt, I.A., Jr. (1961): Design of seawalls and breakwaters, *Trans. of ASCE*, 126, IV, pp. 542-570.

Lesieur, M. and Metais, O. (1996): New trends in large-eddy simulations of turbulence, *Ann. Rev. Fluid Mech.*, Vol. 28, pp. 45-82.

Lin, P. and Liu, P.L.-F. (1998): A numerical study of breaking waves in the surf zone, *J. Fluid Mech.*, Vol. 359, pp. 239-264.

Miller, R.L. (1976): Role of vortices in surf zone prediction: sedimentation and wave forces, Beach and Nearshore Sedimentation, *Soc. Econ. Paleontol. Mineralog.*, Spec. Publ. No. 23, pp. 92-114.

Mizuguchi, M. (1980): An heuristic model of wave height distribution in surf zone, *Proc. 17th Coastal Eng. Conf.*, pp. 278-289.

Nguyen, The Duy and T. Shibayama (1997): A Convection - Diffusion Model for Suspended Sediment in the Surf Zone, *Journal of Geophysical Research*, Oceans, Vol. 102, No. C10, pp. 23169-23186.

Okayasu, A., Shibayama, T. and Horikawa, K. (1988): Vertical variation of undertow in the surf zone, *Proc. of 21st Coastal Eng. Conf.*, ASCE, pp. 478-491.

Okayasu, A., Suzuki, T. and Matsubayashi, Y. (2005): Laboratory experiment and three-dimensional large eddy simulation of wave overtopping on gentle slope seawalls, *Coastal Eng. Journal*, Vol. 47, No. 2-3, pp. 71-89.

Rattanapitikon, W. and Shibayama, T. (2000): Verification and Modification of Breaker Height Formulas, *Coastal Engineering Journal*, JSCE, 42(4), 389-406.

Rattanapitikon, W. and Shibayama, T. (2000): Simple Model for Undertow Profile, *Coastal Engineering Journal*, JSCE, 42(1), 1-30.

Rattanapitikon, W. and Shibayama, T. (1998): Energy Dissipation Model for Regular and Irregular Breaking Waves, *Coastal Engineering Journal*, JSCE, 40(4), 327-346.

Sato, S., Ozaki, M. and Shibayama, T. (1990): Breaking Conditions of Composed and Random Waves, *Coastal Engineering in Japan*, JSCE, 33(2), 133-144.

Sato, S., Isayama, T. and Shibayama, T. (1989): Long-Wave Component in Near-Bottom Velocities under Random Waves on a Gentle Slope, *Coastal Engineering in Japan*, JSCE, 32(2), 149-160.

Schumann, U. (1987): *Direct and Large Eddy Simulation of turbulence – summary of the state of the art 1987*, Lecture series 1987-06, Introduction to the modeling of turbulence, Von Karman Institute for Fluid Dynamics, Rhode Saint Genese, Belgium, pp. 1-36.

Shibayama, T. and Nguyen, T. Duy. (1994): A 2-D Vertical Model for Wave and Current in the Surf Zone Based on the Turbulent Flow Equations, *Coastal Engineering in Japan*, JSCE, 37(1), 41-67.

Smagorinsky, J. (1963): General circulation experiments with the primitive equations, *Mon. Weath. Rev.*, Vol. 91, No. 3, pp. 99-164.

Suzuki, T., Okayasu, A. and Shibayama, T. (2007): A numerical study of intermittent sediment concentration under breaking waves in the surf zone, *Coastal Engineering*, 54, pp. 433-444.

Van de Graaff, J. (1988): *Sediment concentration due to wave action*, Diss., Delft Univ. of Technology.

Van Rijn, L.C. (1984): Sediment pick-up functions, *J. Hydraulic Eng.*, Vol. 110, No. 10, pp. 1494-1502.

Van Rijn, L.C. (1984): Sediment transport, Part II: Suspended load transport, *J. Hydraulic Eng.*, ASCE, Vol. 110, No. 11, pp. 1613-1641.

Watanabe, A. and Maruyama, K. (1986): Numerical modeling of nearshore wave field under combined refraction, diffraction and breaking, *Coastal Eng. in Japan*, Vol. 29, pp. 19-39.

Wijayaratna, N. and Okayasu, A. (2000): DNS of wave transformation, breaking and run-up on sloping beds, *Proc. 4th Int. Conf. on Hydrodynamics*, Vol. 2. IAHR, Yokohama, Japan, pp. 527-532.

Yabe, T., Ishikawa, T., Kadota, Y. and Ikeda, F. (1990): A multidimensional Cubic-Interpolated Pseudoparticle (CIP) method without time splitting technique for hyperbolic equations, *J. Phys. Soc. Japan*, Vol. 59, No. 7, pp. 2301-2304.

Zedler, A.E. and Street, R.L. (2002): Nearshore sediment transport: unearthed by large eddy simulation, *Proc. 28th Int. Conf. Coastal Eng.*, ASCE, pp. 2504-2516.

Chapter 7

Wind Waves

Waves are generated by winds. In order to know the dimensions of generated waves, we must first clarify the energy transmission process from wind to ocean surface.

7.1 Wind Generated Waves

The important parameters in the wave generation process are the wind speed U_{10}, storm duration t_D fetch f and width of wind area B. Figure 7.1 shows a simplified model of a wind generated area. In the figure, U_{10} is the wind speed at 10 m above the water surface, which is frequently used as a standard measure for wind speed. The figure shows a simple box storm coming into the area. In the storm area, the wind velocity U_{10} is between $0 \le t \le t_D$ (t_D: duration of storm).

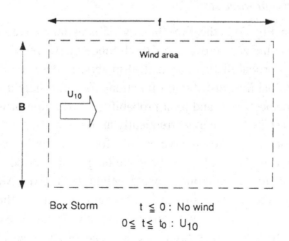

Figure 7.1 A simple model for wind wave.

99

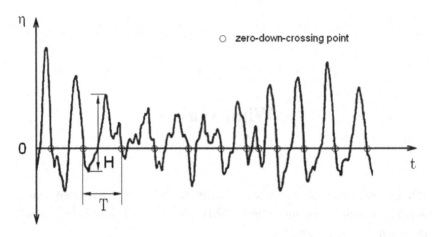

Figure 7.2 Time history of surface elevation.

Under the above condition, as an example, wind waves can be generated. Figure 7.2 shows a typical example of ocean surface elevation, showing an irregular wave history. If we use, for instance, the zero-down-crossing method, individual waves can be defined and the wave height H and wave period T for individual waves can also be determined.

7.1.1 *Individual wave statistics*

As shown in Fig. 7.2, the time history of irregular wave H,T can be determined by the zero-crossing method. Since the wave is irregular, we have to use a probabilistic or statistical method to describe the physical processes. Probability distribution functions for wave height and period, $p(H)$, $p(T)$ respectively, and joint probability distribution of wave height and period $p(H,T)$ are the most frequently used methods.

Classically, representative waves for an irregular wave group were usually used. H_{max}, T_{max} is the wave height and period of the wave corresponding to the maximum wave height in a given wave group. $H_{1/10}, T_{1/10}$ are the average of the heights and periods of the one-tenth highest waves of a given wave group. The significant wave, which is most frequently used, $H_{1/3}, T_{1/3}$ is the average of the one-third highest waves. The mean wave $\bar{H}, \bar{T}$ is the average of all waves in a given wave group.

The probability distribution of wave height has a Rayleigh distribution. The distribution is derived from a theoretical consideration and the result of this consideration agrees with the field data. The distribution is

$$p(H)dH = \frac{\pi}{2}\frac{H}{\bar{H}^2}\exp\left[-\frac{\pi}{4}\left(\frac{H}{\bar{H}}\right)^2\right]dH . \qquad (7.1)$$

According to this distribution, the relationship between representative waves is:

$$H_{1/10} \approx 2.03\bar{H} \approx 1.27 H_{1/3} .$$

7.1.2 *Energy spectrum*

Irregular wave history can be expressed as the sum of various wave components, namely

$$\eta(x, y, t) = \sum_{n=1}^{\infty} a_n \cos(xk_n \cos\theta_n + yk_n \sin\theta_n - \sigma_n t + \varepsilon_n) . \qquad (7.2)$$

Based on this view, the energy spectrum $E(k,\theta)$ can be defined as the wave energy density within the wave number range $(k, k + \delta k)$ and wave direction range $(\theta, \theta + \delta\theta)$. Here

$$\bar{\eta}^2 = \sum_{k=0}^{\infty}\sum_{\theta=0}^{2\pi}\frac{1}{2}a_n^2 = \int_0^{2\pi}\int_0^{\infty} E(k,\theta)\,dk\,d\theta \qquad (7.3)$$

If we limit our consideration to one-dimensional problem and set no distribution to the θ direction, the frequency spectrum can be defined instead of the wave number. For frequency spectrum,

$$\bar{\eta}^2 = \int_0^{\infty} E(f)\,df \qquad (7.4)$$

There are a number of proposed spectrum functions based on theoretical or field observations. Two recommended formulas are (1) Pierson-Moskowitz and (2) Bretscheider-Mitsuyasu. Pierson-Moskowitz's spectrum is

$$E_f(f) = \frac{8.10 \times 10^{-3} g^2}{(2\pi)^4 f^5} \exp\left\{-0.74\left(\frac{g}{2\pi U_{19.5} f}\right)^4\right\} \quad (m^2 \cdot s) \quad (7.5)$$

where $U_{19.5}$: wind velocity at the elevation of 19.5 m from the water surface. Bretschneider - Mitsuyasu's spectrum is

$$E_f(f) = 0.257 H_{1/3}^2 T_{1/3} \times \frac{1}{(T_{1/3} f)^5} \exp\left\{-1.03\left(\frac{1}{T_{1/3} f}\right)^4\right\} \quad (m^2 \cdot s). \quad (7.6)$$

These two formulas are the most frequently used to determine spectrum distribution.

For directional wave spectrum, we simply divide the spectrum into two parts.

$$E(f, \theta) = E(f) \cdot G(\theta; f). \quad (7.7)$$

The function G is the directional distribution function.

As waves propagate towards the shoreline, the shape of the wave spectrum changes. Originally, in the area subjected to the wind, the frequency spectrum has a wide distribution from low to high frequency waves. Because of velocity dispersion, that is when low frequency waves travel faster and high frequency waves travel slower, the width of this spectrum distribution becomes narrower as the waves travel away from the wind-generating region. Also in the high frequency area, energy dissipation due to turbulence is high. As a result, as waves propagate, the width of spectrum becomes narrower. The final form of wave propagation is called swell (see Fig. 7.3).

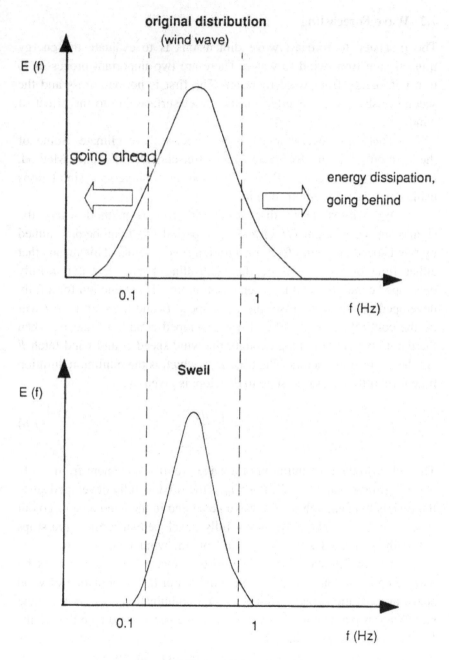

Figure 7.3 Change of energy spectrum.

7.2 Wave Forecasting

The first step to forecast wave dimensions is to evaluate the energy
transmission from wind to water. There are two important processes to
transmit energy from wind to water. The first is normal stress and the
second is shear stress exerted on the water surface due to the effect of
wind.

There are several methods to forecast wave climate. Some of
the methods are: the Sverdrup-Munk-Bretschneider (S-M-B) Method,
Pierson-Neumann-James (P-N-J) Method and numerical simulations
using energy conservation equation.

We will start our discussion with S-M-B method where the
significant wave height $(H_{1/3})$ and wave period $(T_{1/3})$ can be determined
by wind speed and wind fetch or duration time of wind. This means that
either time or space will be the controlling factor. For a non-fully
developed condition, that is, when duration time is not enough for a fully
developed condition to arise, the wind speed U and duration time t will
be the controlling factors. For fully developed condition, that is, when
the duration of time is long enough, the wind speed U and wind fetch F
are the controlling factors. The time $t_{\min}$, which is the minimum duration
time for a fully developed state to develop, is given by,

$$t_{\min} = \int_0^F \frac{dx}{c_g} \tag{7.8}$$

This refers to the time duration of a wave group's movement from $x = 0$
at $t = 0$ till it reaches $x = F$. If $t > t_{\min}$, the wave is fully developed since
the time is long enough for the wave generated in the wind area to go out
from it. If $t < t_{\min}$, the wave is not fully developed since the wind stops
before the generated wave can go out from the wind area.

Figure 7.4 shows the diagram to forecast wave dimensions by
using a combination of wind speed and fetch or the combination of wind
speed and duration time. From the given combinations, we look for the
significant wave height and significant wave period, and take the results
for the lower wave height. P.N.J method includes the effect of frequency
and directional spectrum but will not be described in this book.

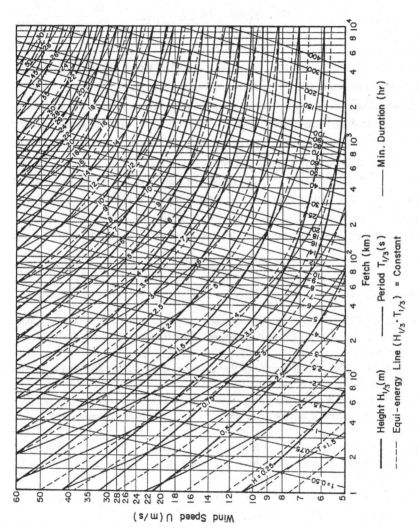

Figure 7.4 Deep water forecasting curves based on SMB method (Handbook of Hydraulic Formulas, JSCE, 1980).

Nowadays, the precise forecast of wind wave development is done by a numerical simulation model. The simulation is based on the following energy conservation equation:

$$\frac{\partial E}{\partial t} + \frac{\partial x}{\partial t}\frac{\partial E}{\partial x} + \frac{\partial k}{\partial t}\frac{\partial E}{\partial k} = S.$$ (7.9)

At the left-hand side, the first term is the time change of energy spectrum. The second term is the convection term and $\frac{\partial x}{\partial t}$ is equal to c_g (group velocity). The third term is the effect of refraction and shoaling. At the right-hand side, the term S is a source term including energy transfer from wind to wave, energy transfer from one wave component to next and energy loss due to wave breaking.

As a conclusion, it is advisable to use SMB method to forecast significant wave dimensions. If the frequency or directional spectrum is necessary, the use of PNJ Method is recommended. When more precise forecasting is required, please use the numerical simulation based on Eq. (7.9).

References

Bretschneider, C.L. (1959): Wave variability and wave spectra for wind generated gravity waves, *U.S. Army Corps of Engineers*, Beach Erosion Board Tech. Memo, No. 118, 192 pp.

Mitsuyasu, H. (1970): On the growth of the spectrum of wind-generated waves, *Coastal Eng. in Japan*, Vol. 13, pp. 1-14.

Pierson, W.J., Neumann, G. and James, R.W. (1955): Observing and forecasting ocean waves by means of wave spectra and statistics, *U.S. Naval Oceanogr.* Office, Pub. No. 603, 284 p.

Pierson, W.J. and Moskowitz, L. (1964): A proposed spectrum form for fully developed wind seas based on the similarity theory of S.A. Kitaigorodskii, *J.G.R.*, 69, No. 24, pp. 5181-5190.

Sverdrup, H.U. and Munk, W.H. (1947): *Wind sea and swell, Theory for relation for forecasting*, U.S. Hdrogr. Office, Washington, Pub. No. 601, 44 p.

Exercise

Problem 7.1 A storm comes to Gulf of Thailand. It is supposed that in the deep water region, 1000 km south from Bangkok (more specifically, the river mouth of Chao Phraya river), there was a constant wind area where the wind speed was 20 m/s, duration time was 5 hours and the fetch of wind area was 200 km.

1. Estimate the significant wave height and significant wave period by using SMB Method.
2. Estimate the time when the wave will come to Chao Phraya River mouth, which is close to Bangkok. Exclude the effect of shoaling in this problem.
3. If we include the effect of shoaling to the estimation in Q.(2), does the time become shorter or longer?

Chapter 8

Wave Induced Currents

There are four possible types of nearshore currents caused by the different physical forces in the water body and the environment. They are the tidal current, the wave induced current, the wind induced current and the sub-stream current due to large scale current. Though these four types of nearshore currents will be mentioned in this chapter, detailed discussion will be limited to the wave induced current.

8.1 Currents in Nearshore Field

The four different types will be explained in this section. The first type is tidal current, whose driving force is influenced by the planetary force balance or gravitational balance of the moon and the sun. Calculations of tidal currents can be performed by using two governing equations: the mass conservation and momentum conservation equations. However, the tidal ranges of the study area have to be specified as a boundary condition. The local topography will also have a strong effect on tidal currents.

The second type is wave induced current. As waves propagate towards the shoreline, they transport mass and momentum in their propagating direction. The transport of mass and momentum causes nearshore current systems.

The third and fourth types are: wind induced current and sub-stream current due to large scale current (such as Kuroshio in East Asia and Gulf of Mexico stream in North America). Wind induced current is significant under storms, but since this phenomenon is more relevant for Disaster Prevention matters we shall discuss it in Chapter 12. The mechanism to generate a large scale current involves an understanding of the global aspect of the wind system of the Earth and the circulation of ocean water, and therefore, will not be included in this book.

Having had a brief overview of the different nearshore currents, we shall now proceed with the description of the mechanisms of wave induced current. In the succeeding sections, we shall discuss the mechanism of this particular current in the two ways in which it is generally described: first by using the mass conservation law, and second by momentum conservation law.

When a water mass is transported shoreward and there is a slope in the water surface (with the area closest to the shore being at a higher level than the offshore area), the slope induces a rip current, which is an offshore directed current. If the wave propagation direction is not perpendicular to the shoreline, the mass transport creates a longshore component and this component becomes part of the longshore current. We can give a qualitative explanation of the above by using the mass conservation equation. If we would like to give a more quantitative description of the nearshore current, it is necessary to use momentum conservation law. From this point, I would like to start our detailed discussion of wave induced current by introducing the concept of radiation stress which is a wave induced stress that generates nearshore currents (momentum flux accompanied by wave propagation).

8.2 Radiation Stress

8.2.1 *The concept of radiation stress*

The concept of radiation stress was introduced by Longuet-Higgins and Stewart (1960) (also see Longuet-Higgins and Stewart, 1964). The radiation stress is, by definition, the average value of the sum ($p_w + \rho V_n^2$) with respect to time, integrated along a vertical plane of unit width. The velocity component perpendicular to the plane is V_n, and ρV_n^2 is the momentum flux. The quantity p_w is the wave pressure from the still water level, i.e. $p_w = p - \rho g z$ where p is the total pressure.

When the plane is vertical and parallel to the wave crest ($V = u$, where u is the horizontal component of water velocity normal to the shoreline), the radiation stress is given by

$$S_{xy} = \frac{1}{T} \int_0^T \int_{-h}^{\eta} (p_w + \rho u^2) \, dz \, dt \qquad (8.1)$$

where $S_{\alpha\beta}$ is a tensor expressing the α-direction flux of β-momentum. We may calculate Eq. (8.1) as

$$\frac{1}{T} \int_0^T \int_{-h}^{\eta} (p_w + \rho u^2) \, dz \, dt = \frac{1}{T} \int_0^T \left\{ \underbrace{\int_{-h}^{\eta} \rho u^2 \, dz}_{I} + \underbrace{\int_{-h}^{0} (p - p_0) \, dz}_{II} + \underbrace{\int_0^{\eta} p \, dz}_{III} \right\} dt$$

$$I + II = \frac{1}{4} \frac{\rho g H^2 kh}{\sinh 2kh}$$

$$III = \frac{1}{16} \rho g H^2.$$

Therefore the radiation stress is

$$S_{xx} = \frac{1}{16} \rho g H^2 \left[1 + 2 \frac{2kh}{\sinh(2kh)} \right]. \qquad (8.2)$$

In shallow water, $S_{xy} = \frac{3}{16} \rho g H^2$, and in deep water, $S_{xy} = \frac{1}{16} \rho g H^2$.

8.2.2 Wave set-up and set-down

This concept is an application of the radiation stress calculation and is important to understand the nature of radiation stress. Consider the momentum balance in a slice of water bounded by a free surface, a gently sloping bottom, and two vertical planes parallel to the wave crest, as shown in Fig. 8.1. The total forces exerted on the planes are the sum of the hydrostatic pressure, radiation stress, and the horizontal

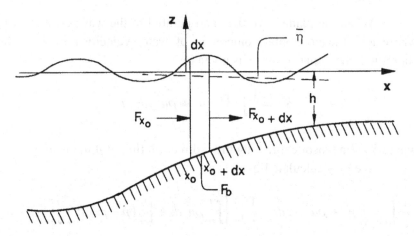

Figure 8.1 The balance of horizontal momentum (after Longuet-Higgins, 1970).

component of the bottom pressure. Momentum balance gives the following relationships,

$$F_{x_0} + F_{x+dx_0} + F_b \cong -\left[\frac{dS_{xx}}{dx} + \rho g(h+\bar{\eta})\frac{d\bar{\eta}}{dx}\right]dx \qquad (8.3)$$

where

$$F_{x_0} = \left[S_{xx} + \frac{1}{2}\rho g(h+\bar{\eta})^2\right]$$

$$F_{x_0+dx_0} = -\left\{S_{xx} + \frac{d}{dx}(S_{xx})dx + \frac{1}{2}\rho g\left[(h+\bar{\eta})^2 + \frac{d}{dx}(h+\bar{\eta})^2 dx\right]\right\}$$

$$F_b = -\left[\rho g(h+\bar{\eta})\frac{d(-h)}{dx}dx\right].$$

One also finds from Eq. (8.3) that

$$\frac{d\bar{\eta}}{dx} = -\frac{1}{\rho g(h+\bar{\eta})}\frac{dS_{xx}}{dx}. \qquad (8.4)$$

Outside the surf zone, $\bar{\eta} \ll h$, therefore

$$\frac{d\bar{\eta}}{dx} \cong -\frac{1}{\rho g h}\frac{dS_{xx}}{dx}.$$ (8.5)

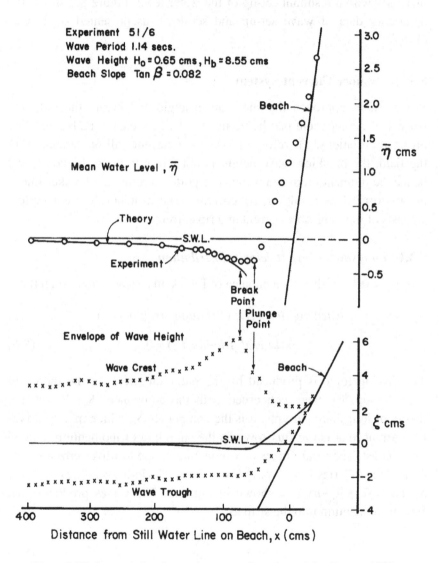

Figure 8.2 Laboratory data for wave set-up and set-down (after Bowen, 1969).

Since S_{xy} increases as the wave moves from deep water to shallow water (to the break point), the mean water level $\bar{\eta}$ decreases, i.e., there is a set-down of the mean water level. After breaking, energy is dissipated and the wave height decreases; therefore S_{xy} decreases and $\bar{\eta}$ increases with a resultant set-up of the water level. Figure 8.2 shows the laboratory data of wave set-up and set-down as presented by Bowen (1969).

8.3 Nearshore Current System

When waves approach the shore at an angle and break, the radiation stress has a component parallel to the coast which cannot be balanced by pressure variations. Therefore, a longshore current will be generated. In the field, the breaking wave height is not uniform along the coast, and hence the momentum of sea water transported through the breaker line is not uniform. This results in rip currents. The nearshore current system consists of the longshore current and rip currents.

8.3.1 *Longshore current velocity distribution*

It is assumed that there is a balance of forces, in a steady state, such that

$$(driving\ force;D_y) + (bottom\ friction;B_y)$$
$$+ (lateral\ friction;L_y) = 0. \tag{8.6}$$

The driving force is produced by the radiation stress. In the theory of wave set-up, we were concerned with the component S_{xx}. But in the theory of longshore currents, it is the component S_{xy} which mainly drives the current, that is the flux towards the shoreline of momentum parallel to the coast. The mean fluxes of y-momentum, due to waves crossing two parallel boundaries $x = x_0$, $x_0 + dx$ of a strip of width dx, are respectively S_{xy} and $(S_{xy} + \frac{\partial S_{xy}}{\partial x} dx)$, as shown in Fig. 8.3. The waves produce a net flux of momentum into the strip as

$$D_y = -\frac{\partial S_{xy}}{\partial x} \tag{8.7}$$

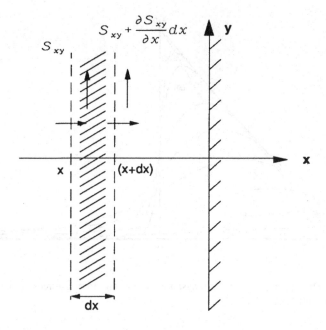

Figure 8.3 The longshore driving force (Longuet-Higgins, 1970).

Using these various assumptions (cf. Longuet-Higgins, 1970) one arrives at,

$$D_y = \frac{5}{4}\alpha^2\rho(gh)^{3/2}\left(-\frac{dh}{dx}\right)\frac{\sin\theta}{c} \qquad (8.8)$$

where θ: wave angle, $\alpha \cong 0.4$, c: phase velocity.

Longuet-Higgins (1970) used the following expressions for the bottom friction and lateral (mixing) friction:

$$\bar{B}_y = \frac{2}{\pi}c_f\rho u_{max}\bar{V} \qquad (8.9)$$

$$\bar{L}_y = \frac{\partial}{\partial x}\left(\varepsilon_\mu\frac{\partial v}{\partial x}\right). \qquad (8.10)$$

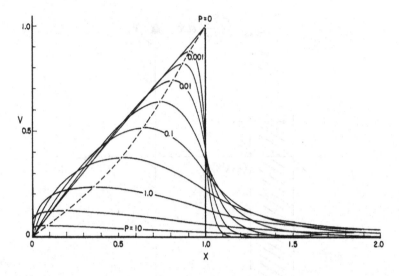

Figure 8.4 Theoretical form of the longshore current, the driving forces are balanced by lateral friction (Longuet-Higgins, 1970).

Here, c_f is a friction coefficient of the order of 0.01, $\bar{V}$: mean longshore velocity, ε_μ : lateral eddy viscosity. As a result, Longuet-Higgins obtained the non-dimensional longshore current velocity distribution given in Fig. 8.4.

8.3.2 *Nearshore circulation*

We start from the following steady equations,

$$u\frac{\partial u}{\partial x}+v\frac{\partial u}{\partial y}=-g\frac{\partial \bar{\eta}}{\partial x}+R_x-\frac{1}{\rho(\bar{\eta}+h)}\left(\frac{\partial S_{xx}}{\partial x}+\frac{\partial S_{yx}}{\partial y}\right)$$

$$u\frac{\partial v}{\partial x}+v\frac{\partial v}{\partial y}=-g\frac{\partial \bar{\eta}}{\partial y}+R_y-\frac{1}{\rho(\bar{\eta}+h)}\left(\frac{\partial S_{xy}}{\partial x}+\frac{\partial S_{yy}}{\partial y}\right)$$

$$(8.11)$$

$$\frac{\partial}{\partial x}[u(\bar{\eta}+h)]+\frac{\partial}{\partial y}[v(\bar{\eta}+h)]=0. \qquad (8.12)$$

where R is a friction term including bottom shear and lateral mixing.

Bowen (1969) assumed that there was no net force outside of the surf zone that might produce a circulation. Inside the surf zone, the driving force is evaluated by using an assumption for the wave height change, for example, $H = \gamma(h + \bar{\eta})$, where γ is a constant. Bowen also made appropriate assumptions for lateral friction and bottom friction. He introduced the transport stream function and obtained a solution which is shown in Fig. 8.5. His solution in the figure shows a closed cell pattern.

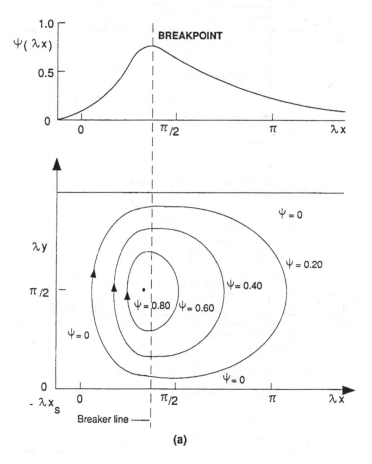

Figure 8.5 The theoretical circulation patter due to waves (after Bowen, 1969). (a) The driving forces are balance by a linear bottom friction (no lateral mixing). (b) The driving forces are balanced by lateral friction.

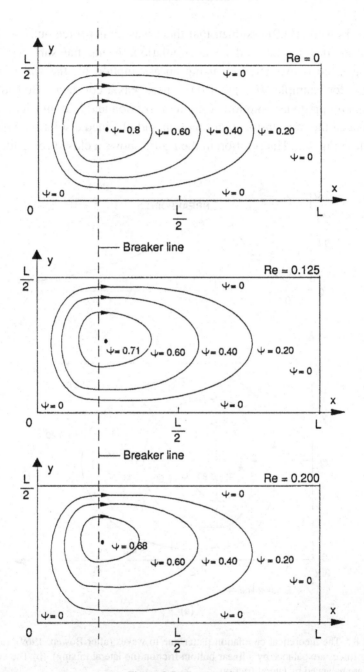

Figure 8.5 (*Continued*)

In a laboratory experiment performed by Okayasu, Hara and Shibayama (1994) nearshore current velocity field are measured as shown in Figure 8.6(1) comparing with their theory. Figure 8.6(2) shows examples of 3-D distribution of nearshore current.

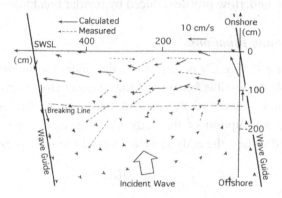

Figure 8.6(1) Measured and calculated depth averaged nearshore current (calculation with momentum flux due to large vortexes) (Okayasu, Hara and Shibayama, 1994).

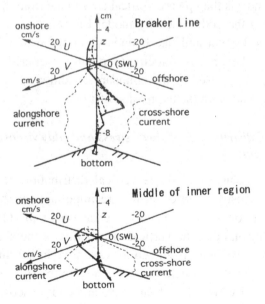

Figure 8.6(2) 3-D distribution of nearshore current (Okayasu, Hara and Shibayama, 1994).

8.4 Undertow Profile (Rattanapitikon and Shibayama, 2000)

Cross-shore time-averaged velocity below wave trough, or undertow, is important in the prediction of the cross-shore suspended sediment transport rate. This section concentrates on the derivation of a model for predicting the undertow profile induced by regular breaking waves.

8.4.1 *Governing equations*

For computing the beach deformation, the undertow model should be kept as simple as possible because of the frequent changing of wave and bottom profiles. Therefore the present undertow profile is calculated based on the assumption of the eddy viscosity model. By considering time-averaged values, the eddy viscosity model can be expressed as

$$\tau = \rho v_t \frac{\partial U}{\partial z} \qquad (8.13)$$

where τ is the time-averaged shear stress, ρ is fluid density, v_t is the time-averaged eddy viscosity coefficient, U is the time-averaged velocity or undertow, and z is the upward vertical coordinate from the bed.

To solve the eddy viscosity model Eq. (8.13), the expression of τ/v_t, should be known and one boundary condition of velocity should be also given. The mean velocity (vertically averaged from bed to wave trough), U_m, is used as the boundary condition of Eq. (8.13). The following part deals with the expression of τ/v_t.

8.4.2 *Vertical distribution of shear stress and eddy viscosity coefficient*

In the eddy viscosity model, the vertical distribution of shear stress, τ, and eddy viscosity coefficient, v_t, are important for the analysis of vertical distribution of the undertow. Okayasu *et al.* (1988) showed through experiments that the vertical distribution of the shear stress and eddy viscosity coefficient, from bed to wave trough, are linear functions of the vertical elevation. Since the turbulence in the surf zone is mainly caused by broken waves, the shear stress and eddy viscosity coefficient may depend on the rate of energy dissipation due to wave breaking. Thus the formula of shear stress, τ, is assumed to be

$$\tau = \rho^{1/3} D_B^{2/3}\left[c_1\frac{z}{d}+c_2 \right] \tag{8.14}$$

where ρ is water density, D_B is the energy dissipation rate of a broken wave, d is the water depth at wave trough, z is vertical coordinate originated from bottom, c_1 and c_2 are the constants.

Rattanapitikon and Shibayama (2000) proposed the following formula to evaluate undertow profile.

$$U = b_1\left(\frac{gH^3}{4Th} \right)^{1/3}\left[b_2\left(\frac{z}{d}-\frac{1}{2} \right)-0.21\left(\ln\frac{z}{d}+1 \right) \right]+U_m \tag{8.15}$$

where U is undertow velocity, U_m is the mean velocity, b_1 and b_2 are the coefficients and are expressed as

$$b_1 = \begin{cases} 0.3+0.7\dfrac{x_b - x}{x_b - x_t} & \text{for transition zone} \\ 1 & \text{for inner zone} \end{cases}$$

$$b_2 = \begin{cases} \dfrac{x_b - x}{x_b - x_t} & \text{for transition zone} \\ 1 & \text{for inner zone} \end{cases}$$

where x is the position in cross-shore direction, x_b is the position at the breaking point, and x_t is the position at the transition point.

8.4.3 *Mean velocity*

The mean velocity, U_m, is assumed to consist of two components, one is due to the wave motion and the other one due to the surface roller (Hansen and Svendsen, 1984).

$$U_m = U_w +U_r \tag{8.16}$$

where U_w is the mean velocity due to wave motion, and U_r is the mean velocity due to the surface roller.

The mean velocity, due to mass transport of wave, M, is

$$U_w = -\frac{M}{\rho h} = -c_3 \frac{B_o g H^2}{ch} \qquad (8.17)$$

where c_3 and B_o are constants. The volume flux below or above the trough level (Q_s) can be written as

$$Q_s = U_w d = c_3 \frac{B_o g H^2}{ch} d . \qquad (8.18)$$

The form of Eq. (8.18) is the same as that of Hansen and Svendsen (1984) for linear shallow water wave.

Mean velocity due to surface roller is given as

$$U_m = -0.76 \frac{B_o g H^2}{ch} - 1.12 b_3 \frac{B_o c H}{h} \qquad (8.19)$$

where b_3 is the coefficient and expressed as

$$b_3 = \begin{cases} 0 & \text{for offshore zone} \\ \dfrac{1/\sqrt{H} - 1/\sqrt{H_b}}{1/\sqrt{H_t} - 1/\sqrt{H_b}} & \text{for transition zone} \\ 1 & \text{for inner zone} \end{cases} \qquad (8.20)$$

where subscript b indicates the value at the breaking point, and subscript t indicates the value at the transition point.

8.4.4 Comparison with experiments

The undertow profiles are computed based on the procedure in 8.4.2 and 8.4.3 and are compared with laboratory data of Okayasu et al. (1988). Figure 8.7 shows these comparisons and indicates how the model gives a reasonably good prediction in the region out of the bottom boundary layer.

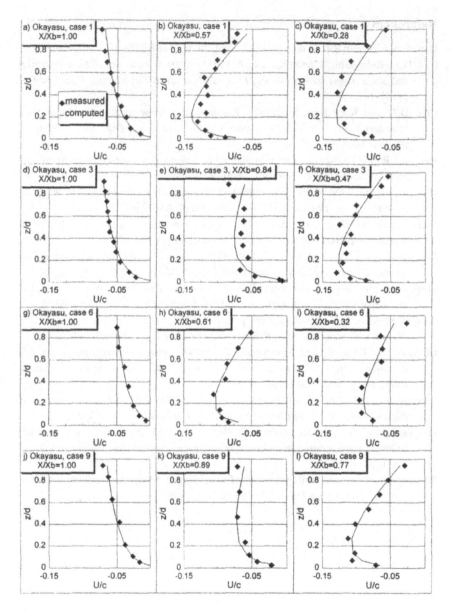

Figure 8.7 Comparison of measured and computed undertow profile (measured data from Okayasu *et al.*, 1988).

References

Bowen, A.J. (1969): Rip currents 1, *J.G.R.*, Vol. 74, pp. 5467-5478.

Cox, D.T., Kobayashi, N. and Okayasu, A. (1994): Vertical variations of fluid velocities and shear stress in surf zones, *Proc. 24th Coastal Engineering Conf.*, ASCE, pp. 98-112.

Cox, D.T. and Kobayashi, N. (1997): Kinematic undertow model with logarithmic boundary layer, *J. of Waterway, Port, Coastal, and Ocean Eng.*, Vol. 123, No. 6, pp. 354-360.

Deigaard, R. and Fredsøe, J. (1991): Modelling of undertow by a one-equation turbulence model, *Coastal Engineering*, No. 15, pp. 431-458.

Hansen, J.B. and Svendsen, I.A. (1984): A theoretical and experimental study of undertow, *Proc. 19th Coastal Engineering Conf.*, ASCE, pp. 2246-2262.

Hansen, J.B. and Svendsen, I.A. (1987): *Surf zone breakers with current*, Nonlinear Water Waves, IUTAM Symposium, Tokyo/Japan, pp. 169-177.

Longuet-Higgins, M.S. and Stewart, R.W. (1960): Changes in the form of short gravity waves on long waves and tidal currents, *J. Fluid. Mech.*, Vol. 8, pp. 565-583.

Longuet-Higgins, M.S. and Stewart, R.W. (1964): Radiation stress in water waves: a physical discussion, with applications, *Deep-Sea Res.*, Vol. 11, pp. 529-562.

Longuet-Higgins, M.S. (1972): Recent progress in the study of longshore currents, in Waves on Beaches edited by R.E. Meyer, Academic Press, pp. 203-248.

Okayasu, A., Hara, K. and Shibayama, T. (1994): Laboratory experiments on 3-D nearshore current and a model with momentum flux by breaking waves, *Coastal Engineering Conf.*, ASCE, pp. 2461-2475.

Okayasu, A., Shibayama, T. and Horikawa, K. (1988): Vertical variation of undertow in the surf zone, *Proc. 21st Coastal Engineering Conf.*, ASCE, pp. 478-491.

Rattanapitikon, W. and Shibayama, T. (1993): Vertical distribution of suspended sediment concentration in and outside surf zone, *Coastal Eng. in Japan*, Vol. 36, No. 1, pp. 49-65.

Rattanapitikon, W. and Shibayama, T. (2000): Simple Model for Undertow Profile, *Coastal Engineering Journal*, JSCE, 42(1), pp. 1-30.

Chapter 9

Wave Forces on Structures

In order to design coastal structures, it is necessary to evaluate wave forces on structures. There are two typical structures employed in coastal construction: breakwaters and cylinders. Cylinders are used for the base of structures such as bridges and breakwaters are typically used in the protection of harbor areas. In this chapter, wave forces on these typical coastal structures will be discussed. Breaking wave pressures will also be included.

9.1 Breakwater

When we construct a rubble mound breakwater, the size of the stone in cover layer should be designed to be stable under high wave attack. This is a problem of force balance exerted on the stone. Figure 9.1 shows the situation. The force balance along the slope with angle α is

$$mg \sin \alpha + (Wave\ Force) = \mu\, mg \cos \alpha \qquad (9.1)$$

where m is mass of stone, μ is a friction coefficient and the mass force in water is proportional to $(\rho_s - \rho_w)d^3 g$. The diameter of stone is d, the density is ρ and subscript s means stone and w means water.

The wave force F_w can be

$$F_w = \frac{1}{2} f_w \rho_w u^2 \times d^2 \times constant . \qquad (9.2)$$

Here f_w is friction factor and we set an assumption that the velocity u can be expressed as (with the assumption of long wave)

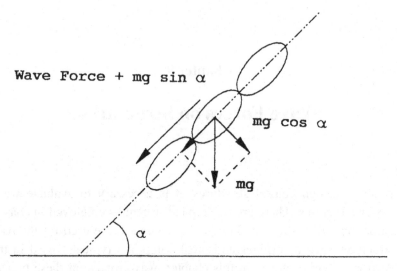

Figure 9.1 Balance of force.

$$u = constant \times \sqrt{gH_B} \qquad (9.3)$$

where H_B is the breaking wave height.

Finally, the design mass of stone M can be calculated as

$$M = \frac{\rho_s H_B^3}{K(\rho_s/\rho_w - 1))^3 (\mu \cos \alpha - \sin \alpha)^3}. \qquad (9.4)$$

Here, the breaking wave height H_B can be changed to the design wave height H_D, and K is a global constant.

Traditionally, Hudson (1959) formula has been used instead of Eq. (9.4). The Hudson formula is

$$M = \frac{\rho_s H_D^3}{k_D(\rho_s/\rho_w - 1))^3 \cot \alpha} \qquad (9.5)$$

where H_D is the local wave height. The value of k_D changes from 2.8 to 4.3 depending on the stone. One of the purpose of using artificial amour units is to increase the value of k_D. For tetrapods, for example, the value of k_D is between 6.6 and 13.6.

9.2 Forces on Cylinder

For coastal and ocean construction, the cylindrical-type structures are frequently employed. The wave forces on a cylinder can be evaluated by Morison's (1950) formula (see Fig. 9.2). The horizontal component of the wave force can be evaluated by the sum of two contributions, drag and inertia.

$$dF = \frac{1}{2}\rho C_D u \,|\,u\,|\,dS + \rho C_M \frac{\partial u}{\partial t} dV \qquad (9.6)$$

where C_D: drag coefficient, dS: projection area to flow direction, C_M: added mass coefficient and dV: volume of element.

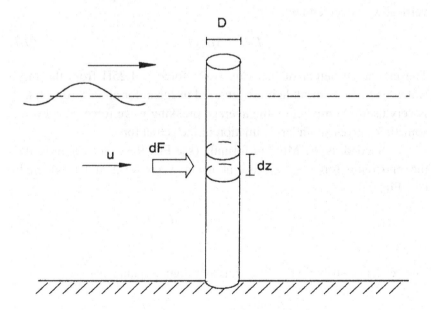

Figure 9.2 Morison formula for wave force to cylinder.

The value for u, $\frac{\partial u}{\partial t}$ can be evaluated by either small amplitude wave theory or finite amplitude wave theory. The drag coefficient C_D is a function of Reynolds number $\frac{UD}{\nu}$ (U: velocity amplitude) and Keulegun-Carpenter (K.C.) number $\frac{UT}{D}$. The value UT is proportional to the horizontal execution of water particle and therefore the physical meaning of the K.C. number is a ratio between water particle excursion and cylinder diameter. The first approximation to value of C_D is 1.2. For the added mass coefficient C_M, the physical meaning of the term is the contribution of potential flow. The first approximation to the value of C_M is 2.0.

9.3 Breaking Wave Pressure

In designing coastal structures, breaking waves approaching a vertical wall are considered to exert the most critical wave force, and should be used when establishing the structural design criteria. There are three practical formulas to evaluate the breaking wave pressure force.

First, is the Hiroi (1919) formula. The breaking wave force F_B on vertical wall is given by

$$F_B = 1.5\rho_w gH .\tag{9.7}$$

The maximum height of breaking wave force is 1.25H from the mean water level. This formula has been used in Japan for almost 70 years. It is very useful in predicting the average breaking wave force. However, it sometimes gives an under-estimation of the actual force.

Second, is the Minikin formula (see Fig. 9.3). According to this, the maximum force p_m appears at mean water level and is given by (see Fig. 9.3),

$$p_m = 102.4\rho_w gd\left(1+\frac{d}{h}\right)\frac{H}{L} .\tag{9.8}$$

The vertical distribution of force has a triangular nature as given by

$$p(y) = \rho_m\left(\frac{H-2|y|}{H}\right)^2 \quad for \quad -\frac{H}{2} < y < \frac{H}{2} .\tag{9.9}$$

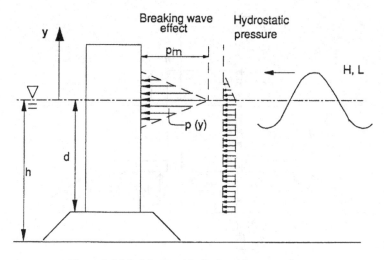

Figure 9.3 Minikin formula for breaking wave force.

The hydrostatic pressure $p_s(y)$ is

$$p_s(y) = \rho_w g \left(\frac{H}{2} - y \right) \quad for \quad y > 0$$

$$= \rho_w g \frac{H}{2} \qquad for \quad y < 0. \qquad (9.10)$$

This formula has been frequently used in U.S.A and also gives a first approximation of breaking wave force. If a more precise estimation is required, there are several formulas which have been recently proposed.

The third and most commonly used method to determine the wave pressure on a vertical breakwater is that formulated by Goda (1974) and successively modified by Takahashi (1994). The model assumes the existence of a trapezoidal pressure distribution along a vertical wall, as shown in Fig. 9.4, regardless of whether the waves are breaking or nonbreaking. In the figure, h denotes the water depth in front of the breakwater, d the depth above the armour layer of the rubble foundation, h' the distance from the design water level to the bottom of the upright section and h_c the crest elevation of the breakwater above the design water level.

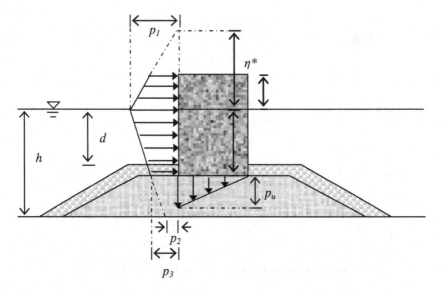

Figure 9.4 Distribution of wave pressure on an upright section of a vertical breakwater.

The elevation to which the wave pressure is exerted (η^*) is given by the formula:

$$\eta^* = 0.75(1 + \cos \beta)H_{max} \qquad (9.11)$$

where β denotes the angle between the direction of wave approach and a line normal to the breakwater and H_{max} is the highest wave in the design sea state (which Goda (1985) recommends should be taken as $1.8H_{1/3}$) The following pressure parameters have to be calculated

$$p_1 = \frac{1}{2}(1 + \cos \beta)(\alpha_1 + \alpha^* \cos^2 \beta)\rho g H_{max} \qquad (9.12)$$

$$p_2 = \frac{p_1}{\cosh(2\pi h/L)} \qquad (9.13)$$

$$p_3 = \alpha_3 p_1 \qquad (9.14)$$

in which

$$\alpha_1 = 0.6 + \frac{1}{2}\left[\frac{4\pi h/L}{\sinh(4\pi h/L)}\right]^2 \tag{9.15}$$

$$\alpha^* = \max(\alpha_2, \alpha_I) \begin{cases} \alpha_2 = \min\left[\dfrac{h_b - d}{3h_b}\left(\dfrac{H_{max}}{d}\right)^2, \dfrac{2d}{H_{max}}\right] \\[4mm] \alpha_I = \alpha_{IH}\alpha_{IB} \end{cases} \tag{9.16}$$

$$\alpha_3 = 1 - \frac{h'}{h}\left[1 - \frac{1}{\cosh(2\pi h/L)}\right] \tag{9.17}$$

where h_b is the water depth at a location at a distance $5H_{1/3}$ seaward of the breakwater, $H_{1/3}$ is the significant incident wave height, L is the wavelength at water depth h, ρ is the density of water and g is the acceleration of gravity.

The factor α^* was originally proposed by Takahashi *et al.* (1994) to estimate the intensity of the impulsive waves and replaced the factor α_2 in the original Goda formula. The factors α_{IH} and α_{IB} were also proposed by Takahashi *et al.* (1994) and can be evaluated using the following sets of equations:

$$\alpha_{IH} = \min(H/2, \ 2.0) \tag{9.18}$$

$$\alpha_{IB} = \begin{cases} \cos \delta_2/\cos \delta_1 & \delta_2 \leq 0 \\ 1/\cosh \delta_1 \cosh^{1/2} \delta_2 & \delta_2 > 0 \end{cases} \tag{9.19}$$

$$\delta_1 = \begin{cases} 20\delta_{11} & :\delta_{11} \leq 0 \\ 15\delta_{11} & :\delta_{11} > 0 \end{cases} \tag{9.20}$$

$$\delta_2 = \begin{cases} 4.9\delta_{22} & :\delta_{22} \leq 0 \\ 3.0\delta_{22} & :\delta_{22} > 0 \end{cases} \tag{9.21}$$

$$\delta_{11} = 0.93\left(\frac{B_C}{L} - 0.12\right) + 0.36\left(0.4 - \frac{d}{h}\right) \qquad (9.22)$$

$$\delta_{22} = -0.36\left(\frac{B_C}{L} - 0.12\right) + 0.93\left(0.4 - \frac{d}{h}\right) \qquad (9.23)$$

where B_c denotes the berm length of the rubble mound foundation.

The buoyancy of the displaced volume of the upright section in still water below the design water level has to be calculated. The uplift pressure acting at the bottom of the caisson is assumed to have a triangular distribution as shown in Fig. 9.4, and the value of the toe pressure P_u is given by the following Equations.

$$p_u = \frac{1}{2}(1 + \cos\beta)\alpha_1\alpha_3\rho g H_{max} . \qquad (9.24)$$

The total wave pressure P at the face of the breakwater and its moment around the bottom on an upright section M_p can be calculated using,

$$P = \frac{1}{2}(p_1 + p_3)h' + \frac{1}{2}(p_1 + p_4)h_c^* \qquad (9.25)$$

$$M_P = \frac{1}{6}(2p_1 + p_3)h'^2 + \frac{1}{2}(p_1 + p_4)h'h_c^* + \frac{1}{6}(p_1 + 2p_4)h_c^{*2} \qquad (9.26)$$

in which

$$p_4 = \begin{cases} p_1(1 - h_c/\eta^*) & : \eta^* > h_c \\ 0 & : \eta^* > h_c \end{cases} \qquad (9.27)$$

$$h_c^* = \min(\eta^*, h_c) . \qquad (9.28)$$

The total uplift pressure U and its momentum around the heel of the caisson M_u can be calculated by using

$$U = \frac{1}{2}p_u B \qquad (9.29)$$

$$M_U = \frac{2}{3}UB \qquad (9.30)$$

where B denotes the width of caisson.

9.4 Breaking Impact Pressure on Vertical Breakwater (Thao and Shibayama, 2007)

Breaking waves on vertical or sloping faced coastal structures produce impact pressures high in magnitude and short in duration, compared with pressures exerted by non-breaking waves. Wave impacts on vertical or sloping breakwaters are one of the most severe and dangerous loads that maritime structures can suffer. The pressures measured are much greater than those expected from the parameters associated with the incident wave (wave height H, water depth h, gravity g and water density ρ). This impulse breaking wave pressure may easily exceed $10\rho g(h+H)$. However, in the past, the short duration wave impact pressures have often been neglected although the pressures can be very high. In recent years, model and prototype tests have been conducted to determine the history and spatial distribution of impulse pressures on smooth seawalls to improve the awareness of wave impact pressure. Thao and Shibayama (2007) proposed a simplified theoretical method to investigate breaking wave impact pressure on a vertical breakwater.

If the advancing wave can simply be considered as a two-dimensional mass of water moving toward a vertical rigid wall with some initial velocity, the impulse momentum relation may be written using the following assumption:

$$\int_0^{t_d} F dt = \int_{v_b}^0 M dv \qquad (9.31)$$

where F is the force, M is the mass of water whose velocity is reduced from breaking velocity v_b before impact to zero in time t_d. Duration time t_d is the time interval for the whole impact process.

Assume the mass M remains constant throughout the impact, and the total force on the wall can be replaced by a pressure P acting over a constant area A

$$\int_0^{t_d} F \, dt = A \int_0^{t_d} p(t) \, dt \tag{9.32}$$

$$\int_{v_b}^{0} M \, dv = \rho \cdot Vol \cdot u_b \tag{9.33}$$

where ρ and Vol are the density and volume, respectively, u_b is maximum velocity of water particle.

The volume of the water can be calculated by using the virtual length of the water mass involved the impact l_v (Weggel and Maxwell, 1970).

$$Vol = l_v A \tag{9.34}$$

l_v can be replaced by $K^* L_b$ where K^* (a function of local wave steepness H_b/L_b) is a dimensionless coefficient (Blackmore and Hewson, 1984). As $L = T \cdot c$ so Eq. (9.34) becomes:

$$Vol = A \cdot K^* \cdot T \cdot u_b \,. \tag{9.35}$$

Substituting Eq. (9.32), (9.33), and (9.35) into Eq. (9.31) gives:

$$\int_0^{t_d} p(t) \, dt = \rho \cdot K^* \cdot T \cdot u_b^2 \,. \tag{9.36}$$

From observation and experiments on wave impacts, a relationship was found between the time interval and impact pressure. In order to simplify the calculation, many researchers generally assumed this relationship to be linear. By assuming a function $f(t_r)$ to perform the non-linearity of the time history of impact pressure, Eq. (9.36) becomes:

$$P_m = f(t_r) \cdot K^* \cdot \rho \cdot T \cdot u_b^2 \tag{9.37}$$

where P_m is maximum impact pressure and t_r is the rise time.

Replacing $f(t_r)K^*$ by λ, which has unit of s^{-1} and is a function of the local wave steepness and rise time (Blackmore and Hewson, 1984), gives:

$$P_m = \lambda \cdot \rho \cdot T \cdot u_b^2. \tag{9.38}$$

The value of λ is investigated and determined through laboratory tests by comparing calculated and measured pressures to obtain a perfect agreement. Thus, the value of parameter λ is identified. The experimental results show that $\lambda = 3$ is suitable for numerical simulations.

The values of velocity of the water particle just in front of the breakwater change throughout impact, and thus the impact pressures will be change accordingly. In order to simulate this process more reasonably, the velocity u_b in Eq. (9.38) is replaced by an instantaneous velocity $u_b(t)$. Therefore, the value of impact pressure can be calculated by

$$p(t) = \lambda \cdot \rho \cdot T \cdot [u_b(t)]^2 \tag{9.39}$$

where $p(t)$ is the time dependent impact pressure, correlating with velocity $u_b(t)$. This velocity is calculated by the 3-D LES model as mentioned in Section 6.3.

Figure 9.5 shows the time history for the impact pressure on the vertical breakwater. The simulation obtains results for the rise times of between 2ms to 17ms, depending on the vertical location along the wall. From this figure, it can be seen that the impact pressure time histories are nonlinear, unlike in the assumptions of previous research in which the time traces are almost triangular.

These time histories show three different stages of impulse pressures for the impact process, as seen in laboratory tests. First of all, the impact pressure increases rapidly from zero to a maximum value. After reaching the highest value, the impact pressure reduces quickly to a certain value that corresponds to the end of the impact component of the wave force. Then it continues to decrease gradually towards zero, ending the impact process. This mechanism is reasonable and is similar to that shown in field surveys and laboratory tests.

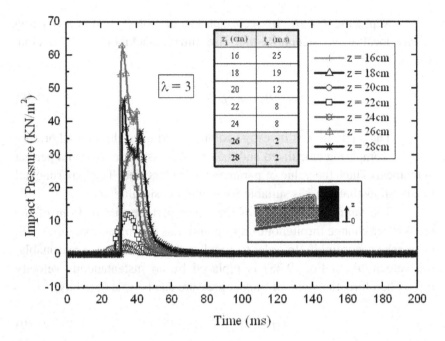

Figure 9.5 Time histories of impact pressures on the vertical breakwater (T = 2s, H_0 = 6cm, h = 16cm, s = 1/20).

The results also show that maximum pressures are inversely proportional to the rise times. The lower the magnitude of the impact pressures, the longer the rise or compression time of the impact process. The maximum pressure 63 KN/m^2 follows a minimum rise time of $2ms$, with longer rise times giving lower impulse pressures. The shape of the wave as it strikes the wall has a strong influence on its impact, and wave pressures with double peaks can be observed for a range of wave conditions, as in the experiments by Blackmore and Hewson (1984).

Figure 9.6 show a series of snapshots of the vertical distribution of computed impact pressures on the wall for the above conditions. From these figures, it can be seen that the region of very high pressures is inside the area where the breaking wave crest hits the wall. In Fig. 9.6, the maximum impact pressures occur above the still water level, regardless of the bottom slope and incident wave properties. This result

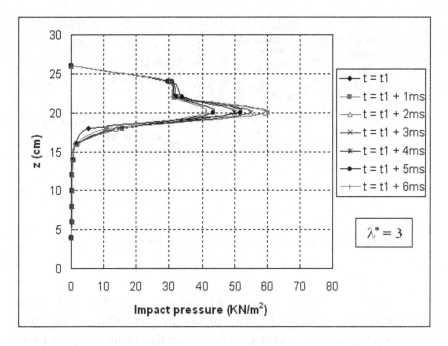

Figure 9.6 Vertical distributions for impact pressure on the vertical breakwater (T = 2s, H_0 = 6cm, h = 16cm, s = 1/20).

is slightly different from the laboratory tests conducted by Kirkgör (1991) and Hattori *et al.* (1994) where the highest pressures on the wall were exerted in the vicinity of the still water level. The reason for this is that the distribution of water mass of the breaking wave from the model is a little above that of the experiments, causing the location of the highest impact pressure to differ slightly.

9.5 Caisson Displacement (Esteban, Takagi and Shibayama, 2007)

Caisson breakwaters can maintain their functionality even if a limited amount of sliding occurs. This sliding is not normally allowed under traditional breakwater design, so Shimosako and Takahashi (2000) proposed a Level 3 design method for caisson breakwaters referred to as the "deformation-based reliability design" that allows for small displacements to occur during the caisson's lifetime.

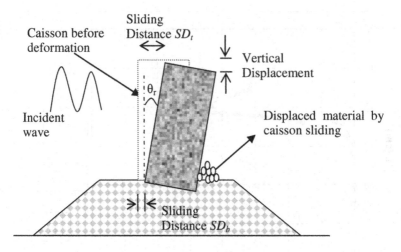

Figure 9.7 Sketch of various parameters.

The European Report PROVERBS (1999) offers a comprehensive review of the state of the art probabilistic breakwater technology. It includes guidance on how to design breakwaters by using a probabilistic approach and recommendations on how to calculate the various forces acting on the breakwater and foundations.

Esteban *et al.* (2007) investigated the long-term deformations of the rubble mound of a caisson breakwater during its working life and proposed a methodology for the calculation of the resistance force caused by the caisson tilting.

As the caisson slides, and due to the fact that a certain part of the caisson becomes imbedded in the foundation due to the vertical displacement at the back, some of the foundation material is displaced. This material accumulates at the back of the caisson and a small but perceptible wedge of material can be observed (see Fig. 9.7). Esteban *el al.* (2007) developed a Monte Carlo simulation to attempt to reproduce these deformations, with the outline of the computational procedure used shown in Fig. 9.8. From this Monte Carlo simulation probability distribution functions for sliding and tilting of the caisson could be computed, which they found to compare reasonably well with the probability distribution functions they obtained through a lengthy series of repetitive laboratory experiments.

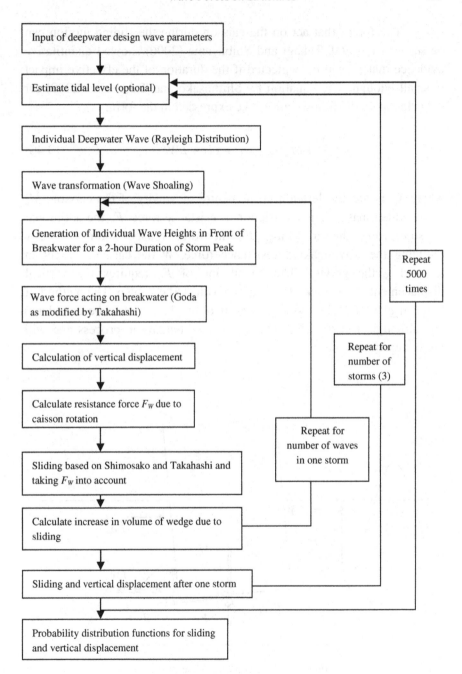

Figure 9.8 Block diagram of computation procedure.

The forces that act on the caisson during the sliding motion can be seen on Fig. 9.9. Takagi and Shibayama (2006) showed quantitative evidence that F_D can be neglected if the duration of the effective impact is small enough. The equation by Shimosako and Takahashi (2000) for the calculation of sliding can thus be expressed in the form:

$$\left(\frac{W}{g} + M_a\right)\ddot{x}_G = P + \mu U - \mu W' - F_W \tag{9.40}$$

where $\ddot{x}_G$ is the acceleration at the centre of gravity of the caisson, M_a is the added mass, F_R is the frictional resistance force, F_w the additional resisting force due to tilting, F_D the force related sliding velocity incliding the wave-making resistance force, W the caisson weight in air and g the gravity. The calculation of F_w requires the vertical displacement at the back of the caisson (tilt) to be known. The method of Esteban *el al.* (2007) allows this vertical displacement to be calculated and although it is straightforward, it is also lengthy in process and will not be reproduced here.

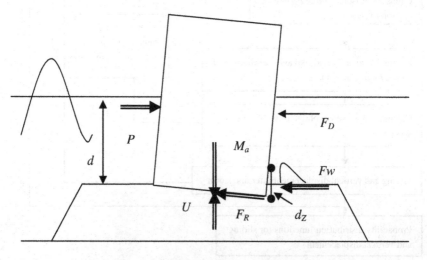

Figure 9.9 Forces acting on the caisson.

References

Blackmore, P.A. and Hewson, P.J. (1984): Experiments on full scale wave impact pressures, *Coastal Eng.*, 8, pp. 331-346.

Esteban, M., Takagi, H. and Shibayama, T. (2007): Improvement in calculation of resistance force on caisson sliding due to tilting, *Coastal Eng. Journal*, Vol. 49, No. 4, pp. 417-441.

Goda, Y. (1974): New wave pressure formulae for composite breakwaters, *Proc 14th Int. Conf. Coastal Eng.*, Copenhagen, ASCE, pp. 1702-1720.

Hattori, M., Arami, A. and Yui, T. (1994): Wave impact pressure on vertical walls under breaking waves of various types, *Coastal Eng.*, 22, pp. 79-114.

Hiroi, I. (1919): On a method of estimating the forces of waves, *J. of College of Eng.*, Tokyo Imperial Univ., Vol. 10, No. 1, pp. 1-19.

Hudson, R.Y. (1959): Laboratory investigation of rubble-mound breakwaters, *Proc. ASCE*, Vol. 85, No. WW3, pp. 93-121.

Kirkgöz, M.S. (1991): Impact pressure of breaking waves on vertical and sloping walls, *Ocean Eng.*, 18, pp. 45-59.

Morison, J.R., O'Brien, M.P., Johnson, J.W. and Schaaf, S.A. (1950): Forces exerted by surface waves on piles, *Petro. Trans. of American Inst. of Mining Eng.*, Vol. 189.

PROVERBS (1996-1999): *Probabilistic design tools for vertical breakwater concept prepared MAST contrast MA53-CT 95-0041*, Leichtweiss Institut fur Wasserben, Braunschweiq, 1999.

Shimosako, K. and Takahashi, S. (2000): Application of expected sliding distance method for composite breakwaters design, *Proc 29th Int. Conf. Coastal Eng.*, ASCE, pp. 1885-1897.

Takagi, H. and Shibayama, T. (2006): A new approach on performance-based design of caisson breakwater in deep water, *Proc. of Coastal Eng.*, JSCE, 53, 901-905 (in Japanese).

Thao, N.D. and Shibayama, T. (2007): Numerical simulation of wave impact pressure on vertical breakwater, Proc. APAC IV, 231-244 (Cd-Rom).

Weggel, J.R. and Maxwell, W.H. (1970): Numerical model for wave pressure distributions, Proc. ASCE, J. Waterw. Harbors Coastal Eng. Div., WW3: 623-642.

Chapter 10

Coastal Sediment Transport

There are two major problems that originate from coastal processes. The first one is erosion and the other is deposition. Erosion occurs when high waves attack shorelines during stormy or monsoon conditions. Deposition, on the other hand, is a problem encountered when waterways or inner port areas are infilled by moving sediments. In this second case it would therefore be necessary for the waterway to be maintained after the construction of the port in order for vessels to continue to use it. Huge amount of dredging is sometimes required each year to maintain water way facilities and port capacities.

An understanding of sediment transport mechanism is, therefore, imperative to solve these two problems. In this chapter, a discussion will be made on two different types of bed materials: sand and mud. Since the transport mechanisms of these materials are distinctly different from each other they shall hence be discussed in separate sections.

10.1 Sand Transport

10.1.1 *General description*

In our study of sediment transport, it is very important to consider the variation changes of coastal lines. To calculate the beach profile change, we will employ the conservation equation for sediment mass which is: (See definition sketch of Fig. 10.1)

$$\frac{\partial h}{\partial t} = -\frac{1}{(1-\Lambda)}\left(\frac{\partial q_x}{\partial x} + \frac{\partial q_y}{\partial y}\right)$$

(10.1)

where h: water depth, t: time, Λ porosity and q_x, q_y: components of time-averaged sediment transport rate in the x-direction and y-directions, respectively. The sediment transport vector $\vec{q}$ can be expressed formally as

$$\vec{q} = \frac{1}{T}\int_0^T \int_{-h(x,y)}^{\eta_1(x,y,t)} c(x,y,z,t)\ \vec{u}_s(x,y,z,t)dzdt \qquad (10.2)$$

where T: wave period, η_1: water surface elevation, c: volumetric concentration of moving sediment, and $\vec{u}_s$: sediment velocity vector.

In order to calculate the sediment transport rate by Eq. (10.2), the instantaneous values of sediment concentration and of sediment velocity must be known. In a first-principle treatment, the momentum equation of individual sediment particles can be used. The momentum conservation equation is, (according to Hinze, 1975 pp. 463),

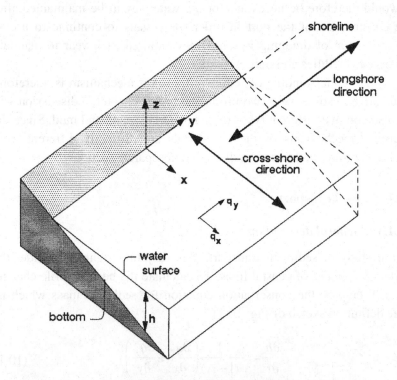

Figure 10.1 Definition sketch giving basic notation.

for laminar flow:

$$\frac{\pi d^3}{6} \rho_s \frac{d\vec{u}_s'}{dt} = \frac{\pi d^3}{6} \rho_f \frac{d\vec{u}_f'}{dt}$$

$\underbrace{\qquad\qquad\qquad}$ $\underbrace{\qquad\qquad}$

Force to accelerate   Pressure
sand particle   gradient

$$-\frac{1}{2}\frac{\pi d^3}{6} \rho_f \left(\frac{d\vec{u}_s'}{dt} - \frac{d\vec{u}_f'}{dt} \right) - 3\pi v \rho_f \, d\,(\vec{u}_s' - \vec{u}_f')$$

Force to accelerate   Viscous force
added mass of particle

$$+\frac{3}{2} \rho_f \sqrt{\pi v} d^2 \int_{t_0}^{t} \frac{\dfrac{d\vec{u}_s'}{dt_1} - \dfrac{d\vec{u}_f'}{dt_1}}{\sqrt{t - t_1}} \, dt_1$$

Basset term

$$+\frac{\pi d^3}{6} (\rho_s - \rho_f) G + F$$

Gravity   External potential   (10.3)
force

where the subscript *s* denotes sediment, the subscript *f* denotes fluid, *d* is the sediment diameter, ρ is the density, $\vec{u}'$ is the Lagrangian velocity vector, and *v* is the kinematic viscosity of fluid.

for turbulent flow:

$$\frac{\pi d^3}{6} \rho_s \frac{d\vec{u}_s'}{dt} = \frac{\pi d^3}{6} \rho_f \frac{d\vec{u}_f'}{dt} - C_M \frac{\pi d^3}{6} \rho_f \left(\frac{d\vec{u}_s'}{dt} - \frac{d\vec{u}_f'}{dt} \right)$$

Force to accelerate   Pressure   Force to accelerate
sand of particle   gradient   added mass of particle

$$-C_D \frac{d^2}{8} \pi \rho_f (\vec{u}_s' - \vec{u}_f') | (\vec{u}_s' - \vec{u}_f') |$$

Viscous force

$$+\frac{3}{2}\rho_f\sqrt{\pi(v+v_e)}\,d^2\underbrace{\int_{t_0}^{t}\frac{\dfrac{d\vec{u}_s'}{dt}-\dfrac{d\vec{u}_f'}{dt}}{\sqrt{t-t_1}}\,dt_1}_{\text{Basset term}}$$

$$+\underbrace{\frac{\pi d^3}{6}(\rho_s-\rho_f)G}_{\text{Gravity}}+\underbrace{F}_{\substack{\text{External potential}\\\text{force}}}$$

(10.4)

where C_D is the drag coefficient, C_M is the added mass coefficient and v_e is the eddy viscosity of fluid.

If the right side of Eq. (10.3) or (10.4) were evaluated, we could calculate the behaviour of individual sediment particles. Then the values of c and $\vec{u}_s$ in Eq. (10.2) could be found by summarizing individual movement. However, it is unrealistic or even impossible to evaluate the right sides of Eq. (10.3) and (10.4) with good accuracy at the present time. Therefore, for our engineering purposes, we will make efforts to solve q in Eq. (10.2) by means of a physical and a more general description of the sediment movement pattern.

The most useful and most frequently used parameter to describe sand motion is the Shields parameter. The Shields parameter ψ_m, which is the non-dimensional bed shear stress, is expressed in terms of the maximum value of the near bottom water velocity, and is defined by

$$\psi_m=\frac{f_w u_b^2}{2(s-1)gd}$$

(10.5)

where f_w: Jonsson's (1966) wave friction factor, u_b: the maximum value of the near bottom velocity, s: the sediment specific gravity, g: the gravitational acceleration, and d: the sediment diameter. When the friction factor, f_w, is calculated, the bottom roughness is taken to be equal to the sediment diameter. The physical meaning of the shields parameters is the ratio between the driving force on sediment particle to the stabilizing force. If we suppose a steady state condition for Eq. (10.4), we may get

$$C_D \frac{d^2}{8} \pi \rho_f (\bar{u}'_f - \bar{u}'_s)^2 + (\rho_s - \rho_f)\bar{g}\frac{\pi d^3}{6} = 0. \qquad (10.6)$$

The Shields parameter means the ratio of these two terms, drag term (driving force) and gravitational term (stabilizing force).

10.1.2 Cross-shore transport formula

In this section, a power model will be introduced. (The discussions until Eq. (10.13) is based on the lecture of Prof. A. Watanabe in 1979). In the power model, the amount of moving sand mass in water is assumed to be proportional to the bottom shear stress, τ_b. Also the velocity of the sand particle is proportional to the water velocity near the bottom, u_b. Since the flux of sand is the product of the amount of moving sand and sand velocity, the mass transport rate, i_s is

$$i_s = N\frac{\pi}{6}(\rho_s - \rho_f)d^3 \times C_1 u_b$$

$$= C_1 C_2 \tau_b u_b \qquad (10.7)$$

where N: number of sand particles moving per unit area and C_1 and C_2 are empirical constants. If we exchange mass to volume, the volumetric transport rate q_S is

$$q_s = \frac{i_s}{\rho_s - \rho_f} = C_1 C_2 \frac{\tau_b u_b}{\rho_s - \rho_f}. \qquad (10.8)$$

Here we use non-dimensional sediment transport rate ϕ by using fall velocity w_0 and sand diameter d as

$$\phi = \frac{q_s}{w_0 d}. \qquad (10.9)$$

The fall velocity w_0 can be expressed by Stokes law (for low Reynolds number) as,

$$w_0 = \sqrt{\frac{4}{3}\frac{(s-1)gd}{C_D}}. \qquad (10.10)$$

Inserting Eq. (10.8) and Eq. (10.10) into Eq. (10.9), and using

$$u_b = \sqrt{\frac{2\tau_b}{\rho_f f_w}}$$

we get

$$\phi = constant \times \left(\frac{\tau_b}{(\rho_s - \rho_f)gd} \right)^{3/2}. \qquad (10.11)$$

From Eq. (3.2.5), Shields parameter is

$$\frac{\tau_b}{(\rho_f - \rho_s)gd} = \psi \qquad (10.12)$$

and

$$\Phi = constant \times \psi^{3/2}. \qquad (10.13)$$

As indicated above, if we set the assumption that the amount of moving sand mass is proportional to the bottom shear stress, we get (Eq. 10.13). Now, let us set an assumption that the amount of moving sand mass is proportional to the order of the 2.5th power of the bottom shear stress. With some simplification and by determining constants with the use of laboratory results, the formula obtained is

$$\Phi(t) = 40\psi^3(t). \qquad (10.14)$$

This formula is frequently used and is referred to as the Einstein-Brown type formula. The formula gives an instantaneous value of transport rate. For wave motion, the transport formula will be applied to evaluate the transport formula averaged over a half-wave cycle. In order to evaluate the net sand transport rate, we divide the time history of the near-bottom velocity into half-waves by using the zero-crossing method and apply the transport rate formula to each half wave.

The transport formula averaged over a half-wave is given as follows. When sediment particles are accelerated, the transport rate can be estimated by Eq. (10.14). Assuming a sinusoidal velocity change, $u(t) = u_b \sin \sigma t$, and also assuming the quasi-steady state, we get

$$\Phi(t) = 40\psi_m^3 \sin^6 \sigma t \quad for \quad t_c < t < T/4 \tag{10.15}$$

in which ψ_m is the maximum value of Shields parameter in half a wave cycle, and $\sigma = 2\pi/T$, where T: wave period and t_c: time when a sediment particle starts to move. For the time interval after a volume of moving sediment attains its maximum volume, we make an assumption that the sediment particles once they are placed in motion will not stop during the half period as stated before. The bed load rate is the product of the volume of moving sediment and its velocity. According to the assumption, the volume of moving sediment does not change, but its velocity changes with the fluid velocity. Then we have

$$\Phi(t) = \Phi_m u(t)/u_b \quad for \quad T/4 < t < T/2 \tag{10.16}$$

in which Φ_m is the maximum value of the transport rate i.e., $40\psi_m^3$. Assuming a sinusoidal velocity change,

$$\Phi(t) = 40\psi_m^3 \sin \sigma t \quad for \quad T/4 < t < T/2. \tag{10.17}$$

The transport rate averaged over the first half wave period is thus estimated via

$$\Phi = \frac{2\left(\int_{t_c}^{T/4} 40\psi_m^3 \sin^6 \sigma t \ dt + \int_{T/4}^{T/2} 40\psi_m^3 \sin \sigma t \ dt \right)}{T} \tag{10.18}$$

$$\Phi = C_1 \psi_m^3 .$$

The coefficient C_1 is a function of the ratio of the maximum value of the Shields parameter ψ_m and the critical value of the Shields parameter which corresponds to initial sand movement ψ_c, i.e. ψ_m/ψ_c. Figure 10.2

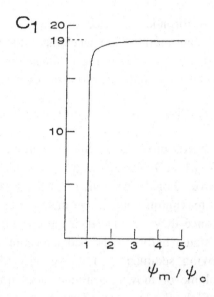

Figure 10.2 The relation between the coefficient C_1 and ψ_m/ψ_c.

shows the relationship between C_1 and ψ_m/ψ_c. If ψ_m is greater than two times ψ_c, the coefficient C_1 was found to be well approximated by the value of 19 (Fig. 10.2). Under such conditions we have (Shibayama-Horikawa, 1980 formula)

$$\bar{\Phi} = 19\psi_m^3 .$$
(10.19)

If we apply Eq. (10.15) to the whole half period and take the time average, we get the following Madsen-Grant (1977) formula.

$$\bar{\Phi} = 12.5\psi_m^3 .$$

10.1.3 *Longshore transport formula*

In the evaluation of the longshore transport rate, we will use the power model wherein the mass of moving sand in water is assumed to be proportional to the bed shear stress under wave-current condition. The velocity of moving sand is proportional to the longshore velocity, and as a result the total transport rate I_e for longshore direction is

$$I_e = constant \times \tau \, \overline{v}_l \, x_B \tag{10.20}$$

where τ: representative shear stress, $\overline{v}_l$ averaged longshore current velocity, x_B: width of surf zone.

Based on the power model, there are two formulas frequently used for practical problems. The first is the Komar and Inman (1970) formula,

$$I_e = 0.77 P_e \tag{10.21}$$

where I_e: immersed weight, $P_e = (EC_n)_B \cos \alpha_B \sin \alpha_B$, subscript B denotes values at breaking point, α is wave angle.

The second is Savage's (1962) empirical formula,

$$Q_l = \alpha P_l \tag{10.22}$$

where Q_l, the volumetric transport rate is equal to $\dfrac{l_e}{(1-\Delta)(\rho_s - \rho_f)g}$ and the value α is equal to 0.217 (m^3/t) as a first approximation. The value α should be evaluated for each individual beach based on field measurements.

10.1.4 *Initiation of sand transport*

It is an important problem to determine what the prevailing condition is when sand particles start to move. This problem is particularly evident in designing a jetty to stop longshore sand transport. In order to come out with reasonable design criteria, it is necessary to evaluate at what depth the sand will start to move.

Madsen and Grant (1976) showed that the critical value of the shields parameter corresponding to the initiation of sediment movement could be expressed as a function of a quantity S. The quantity S was used in place of the boundary Reynolds number, and it depends only on the sediment and fluid properties as shown below:

$$S = \frac{d}{4v}\sqrt{(s-1)gd} \tag{10.23}$$

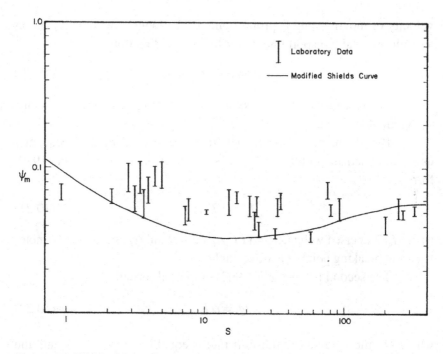

Figure 10.3 Shields diagram for the initiation of sediment movement (after Madsen and Grant, 1976).

where v: the kinematic viscosity of the fluid, and s: specific gravity of sand.

Figure 10.3 shows the result given by Madsen and Grant. For sediment particles of specific gravity 2.65, diameter 0.2 to 2.0 mm and the parameters S, 2.8 to 90 (common values for natural beaches), the critical value of the Shields parameter varies from 0.03 to 0.08 if we take the modified Shields curve in the figure.

10.1.5 *Suspended sediment concentration in and outside the surf zone (Jayaratne and Shibayama, 2004)*

Sediment suspension usually occurs in and outside the surf zone due to vortexes generated by sand ripples, moving of the bottom layer with high bed shear stresses and turbulence generated by wave breaking. A set of predictive models of suspended sediment concentration over rippled seabed, sheet flow layer and under breaking agitation was derived recently by Jayaratne and Shibayama (2004) with the help of small and large-scale laboratory data. A good correlation can be seen in the predicted and measured values of suspended sediment concentration.

a) Suspension due to Vortexes Generated by Sand Ripples

The prediction of the reference concentration, c_r, over rippled seabed is an important task before computing the concentration profile throughout the water column. With the help of a power model, c_r can be written in the following way.

$$c_r = (\psi - \psi_c)\frac{k_{10}w_s d}{f_1\sqrt{(s-1)gd}} \qquad (10.24)$$

where ψ is the Shield's parameter, ψ_c is the critical Shield's parameter, w_s is the settling velocity of sand, d is the grain diameter, s is the specific gravity of sand, g is the gravitational acceleration, k_{10} is a constant and f_1 is a function of ripple dimension (η and λ), grain size (d) and flow properties (w_s and fluid kinematic viscosity, v).

Using the best-fit technique and considering 48 different rippled bed cases, f_1/k_{10} was evaluated and finally Eq. (10.25) was established for c_r to predict at the level of $\eta/2$ above the ripple crest (see Fig. 10.4(1)).

$$c_r = \frac{10(\psi - 0.05)v}{\sqrt{(s-1)gd}\,(\eta/2)} \qquad (10.25)$$

where η is the ripple height and λ is the ripple length.

By considering the previous diffusion coefficient equations of Nielsen (1988) and Shibayama and Rattanapitikon (1993), Eq. (10.26) was proposed for the diffusion coefficient, ε_r, with the help of 100 laboratory data over rippled bed conditions (see Fig. 10.4(2)). It can be seen that diffusion under this condition is mainly governed by η and,

$$\varepsilon_r = 0.27 u_{*wc} A_b \left(\frac{w_s}{u_{*wc}} \right)^2 \left(\frac{\eta}{d} \right)^{0.1} \left(\frac{\lambda}{d} \right)^{0.25} d_*^{-1.5} \qquad (10.26)$$

where u_{*wc} is the shear velocity, A_b is the orbital amplitude of water particle and $d_* = d(sg/v^2)^{1/3}$.

After integrating the steady diffusion equation and substituting values of c_r and ε_r, the concentration profiles under rippled bed fit well with an exponential form calculated by Eqs. (10.25) and (10.26). Figure 10.4(3) shows the comparison with laboratory data.

b) Suspension over Sheet Flow Layer

By selecting suitable constant parameters and changing the reference level in Eq. (10.25), Eq. (10.27) was proposed for the reference concentration, c_s, over sheet flow layer at the level of $25d$ above the mean bed level with the help of 12 laboratory data sets over sheet flow conditions (see Fig. 10.4(4)). Similarly, by considering the diffusion equation proposed over a rippled seabed, Eq. (10.26), and analyzing a total of 10 different cases and using the best-fit technique, Eq. (10.28) was proposed for the calculation of the diffusion coefficient, ε_s (see Fig. 10.4(5)). The main feature in Eq. (10.28) is the dominancy of bed shear velocity with the reciprocal of the ratio w_s/u'_{*wc} and the absence of ripple dimension (η and λ). As in the rippled bed case, the measured and predicted values of suspended sediment concentration are well predicted by an exponential equation (see Fig. 10.4(6)).

$$c_s = \frac{2.75(\psi - 0.05)v}{\sqrt{(s-1)gd}\,(25d)} \qquad (10.27)$$

$$\varepsilon_s = 0.012 u'_{*wc} A_b \left(\frac{u'_{*wc}}{w_s} \right)^2 d_*^{-1.5} \qquad (10.28)$$

where u'_{*wc} is the bed shear velocity under sheet flow condition.

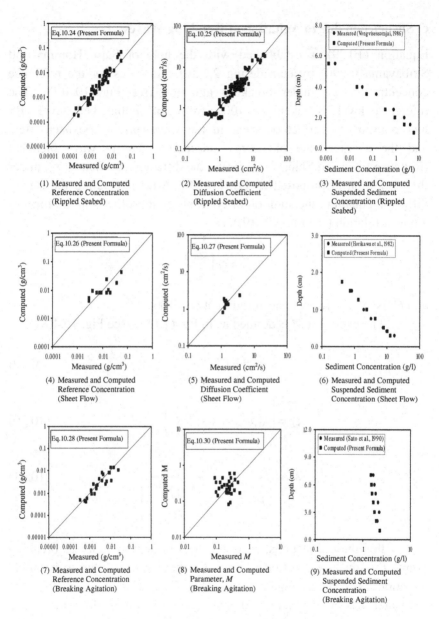

Figure 10.4 Examination of formulas.

c) Suspension due to Agitation of Breaking Waves

Equation (10.29) is established with the help of Sato, Homma and Shibayama (1990) by considering 23 different cases for the reference concentration, c_b, under breaking agitation (see Fig. 10.4(7)). The reference level is taken as $100d$ from the initial bed level. By incorporating the effect of shear in the wave-current coexistent field and the average rate of energy dissipation due to wave breaking (Rattanapitikon and Shibayama, 1998), the diffusion coefficient, ε_b, under this condition is proposed as Eq. (10.30). After integrating the steady diffusion equation, the solution of concentration profile is in the form of a power relation (see Fig. 10.4(9)), as

$$c_{(z)} = c_b \left(\frac{z_0}{z} \right)^M$$

and C_b is the reference concentration at $z_0 = 100d$.

The exponent M is defined as in Eq. (10.31) (see Fig. 10.4(8)).

$$c_b = 10^{-9} gT \frac{\hat{u}_b^{2.3}}{w_s^{3.3}} \tag{10.29}$$

$$\varepsilon_b = \left[k_{11} u''_{*wc} + k_{12} \left(\frac{D_B}{\rho} \right)^{1/3} \right] z \tag{10.30}$$

$$M = \frac{w_s}{\left[k_{11} u''_{*wc} + k_{12} \left(\frac{D_B}{\rho} \right)^{1/3} \right]}. \tag{10.31}$$

For these equations, $\hat{u}_b$ is the near-bottom velocity at breaking point, T is the wave period, u''_{*wc} is the shear velocity due to breaking agitation, D_B is the average rate of energy dissipation, ρ is the density of sea water, z is the vertical coordinate measured from the bottom, k_{11} and k_{12} are constants found from the best-fit technique for different breaker types such as spilling and plunging breakers with 29 different data sets.

The bottom reference concentration and diffusion coefficient over rippled seabed, sheet flow layer and under breaking agitation are evaluated separately using large number of published data. A total evaluation method of suspended sediment in and outside the surf zone is suggested by integrating each component.

10.2   Modelling of Time-Dependent Sand Transport at the Bottom Boundary Layer (Shibayama and Nistor, 1996)

A model was developed by Shibayama and Duy (1994) to simulate the temporal and spatial distributions of suspended sediment in the surf zone (see Section 6.2). The model is solved numerically in a time domain and in a two-dimensional vertical plane. At each time step of computation, the upper moving boundary (water surface) and the velocity field are determined through the solution of a hydrodynamic model which is based on the Reynolds equation of motion. The sediment concentration field is then determined by solving the convection-diffusion equation of the sediment mass. The comparisons with laboratory data of various experimental conditions show reasonable agreements for the simulated wave field and the concentration field in the surf zone.

10.2.1   The turbulent convection-diffusion equation for the sand concentration field

In the bottom boundary layer, intensive concentrations of moving sand have been observed during laboratory as well as field experiments. For this area, the characteristics of the sand transport are strongly influenced by the turbulence induced by the flow field. The physical processes related to the particle movement in the flow are described by the turbulent convection-diffusion equation. The Reynolds decomposition approach was applied for the present sand transport model so that the equi-phase mean values of the sand concentration field can be obtained. The equation is partitioned in such a way that particles are advected by the combination of vertical and horizontal velocities of the fluid and the sand particle settling velocity and diffused by the effects of turbulence.

The two-dimensional convection-diffusion equation is expressed as

$$\frac{\partial c}{\partial t} + \frac{\partial (uc)}{\partial x} + \frac{\partial (wc)}{\partial z} = \frac{\partial (cw_s)}{\partial z} + \frac{\partial}{\partial x}\left(\varepsilon_{sx}\frac{\partial c}{\partial x}\right) + \frac{\partial}{\partial z}\left(\varepsilon_{sz}\frac{\partial c}{\partial z}\right) \quad (10.32)$$

where c: the equi-phase mean sand concentration; u, w: the equi-phase mean velocity components calculated in the hydrodynamic module for the bottom boundary layer, w_s: the mean fall velocity of sand, ε_{sx} and ε_{sz}: the turbulent diffusion coefficients corresponding to the two axis of coordinates. However, for the present model, it was assumed that the coefficients of turbulent diffusion, ε_{sx} and ε_{sz} are equal,

$$\varepsilon_{sx} = \varepsilon_{sz} = \varepsilon_s \tag{10.33}$$

so that Eq. (10.32) can be rewritten as

$$\frac{\partial c}{\partial t} + \frac{\partial(uc)}{\partial x} + \frac{\partial(wc)}{\partial z} = \frac{\partial(cw_s)}{\partial z} + \frac{\partial}{\partial x}\left(\varepsilon_s \frac{\partial c}{\partial x}\right) + \frac{\partial}{\partial z}\left(\varepsilon_s \frac{\partial c}{\partial z}\right). \tag{10.34}$$

Another simplification is the fact that the sand particle settling velocity depends on the sand particle characteristics and is assumed constant throughout the thickness of the boundary layer.

In order to be able to simulate a sloping bottom, the initial system of coordinates (x, z) is transformed into the (α, γ) system by means of a similar procedure to the one of Daubert *et al.* (1982). Figure 10.5 shows the details of the co-ordinates transformation.

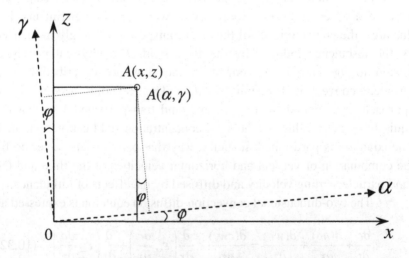

Figure 10.5 Transformation of the coordinates system.

10.2.2 *Formulation of boundary conditions*

Boundary conditions have to be specified for the calculation of both the hydrodynamic field and for the sand concentration field. In order to get the boundary layer flow, the model requires the time-dependent values of the flow field at the upper edge of the bottom boundary layer. These values are calculated using the hydrodynamic model developed by Duy and Shibayama (1997) which employs the 2DV Navier-Stokes equations in their Reynolds averaged form. Therefore, the boundary conditions for the velocity calculation are

$$u = 0 \text{ and } w = 0 \quad \text{for} \quad z = 0 \tag{10.35}$$

$$u = u_o \text{ and } w = w_o \quad \text{for} \quad z = \delta \tag{10.36}$$

where δ: the thickness of the bottom boundary layer.

For the offshore lateral boundary, as well as for the onshore one, the velocity profile is assumed to follow a logarithmic distribution throughout the boundary layer thickness. For the horizontal velocity, u, the previously mentioned distribution is considered as

$$u = u_o \ln(Az + B) = u_o \ln\left(\frac{e-1}{\delta} z + 1\right) \tag{10.37}$$

where constants A and B have been determined by using the upper and bottom boundary conditions for the boundary layer velocities (the constant e is the natural logarithmic base). The boundary conditions for the sand concentration calculation are given at the upper, bottom and lateral boundaries. In the case of the bottom boundary condition, the definition of concentration has an intricate pattern due to the fluid-particle and particle-particle interactions, which are difficult to define. During the last few years, researchers have proposed different conditions for the reference concentration. These conditions reflect both the values of the reference concentration and its vertical location. Different concepts were applied in the past such as the one of the "pick-up rate" proposed by Fredsøe and Deigaard (1992), in which the near-bed concentration is expressed by a gradient boundary condition and assumed to be dependent on the instantaneous and critical value of the Shields parameter and on the sand characteristics.

The reference concentration is given at a certain distance from the bed, which is related to the bottom roughness. Still, for the case of the bottom boundary layer, the laboratory data reveal that the movement of sand is present at elevations which are lower than the level of reference concentration. It is physically intuitive to assume that the reference concentration is almost equal to the maximum concentration of immobile and immersed sand. Therefore, for the sediment module, the bottom boundary condition is

$$c = c_{\max} \quad \text{for} \quad z = 0 \tag{10.38}$$

where $c_{\max}$: the maximum concentration of immobile and immersed sand. The value of the reference concentration for the present model does not need to include the effect of breaking or non-breaking waves. The reason is that $c_{\max}$ is the value of the maximum concentration of *non-moving* sand. For the present section, the maximum value of sand concentration was taken to be equal to the maximum *measured* concentration of sand recorded at the elevation where the sand bed was completely immobile, with no visible movement of sand particles (laboratory experiment of Ribberink and Al-Salem, 1992). The value of the maximum concentration therefore depends only on sand characteristics (particle diameter, natural porosity and degree of compaction). Asano (1990) also successfully showed the validity of a similar assumption.

The reference elevation for the maximum concentration is defined as the elevation of the experimentally observed undisturbed sand bed during the oscillatory motion. This reference level is located under the static bed level, that is, the initial bed level, prior to the start of the oscillatory movement. Choosing such a boundary condition is expected to simulate in a realistic manner the physical aspects of sand movement phenomena after the initiation of the sheet-flow regime. Figure 10.6 shows a schematic view of the boundary condition.

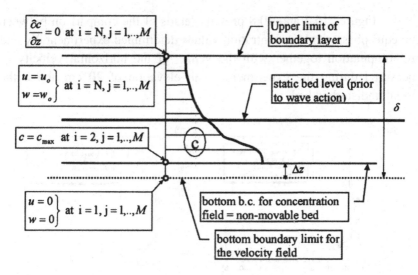

$\frac{\partial c}{\partial z} = 0$ at $i = N, j = 1,..,M$

Upper limit of boundary layer

$\left. \begin{array}{l} u = u_o \\ w = w_o \end{array} \right\}$ at $i = N, j = 1,..,M$

static bed level (prior to wave action)

δ

$c = c_{max}$ at $i = 2, j = 1,..,M$

c

Δz

$\left. \begin{array}{l} u = 0 \\ w = 0 \end{array} \right\}$ at $i = 1, j = 1,..,M$

bottom b.c. for concentration field = non-movable bed

bottom boundary limit for the velocity field

Figure 10.6 Boundary conditions for the hydrodynamical model and sand concentration model inside the bottom boundary layer.

10.2.3 Sand concentration within the bottom boundary layer-comparison with laboratory data

The two modules, hydrodynamic and sand transport, are now coupled to get the equi-phase mean values of sand concentrations inside the boundary layer. For the case of time-dependent suspended sand concentration, many researchers have already performed detailed measurements. However, in sheet flow conditions, measurements in the area close to the bed have rarely been performed and reported. Ribberink and Al-Salem (1992) performed three sets of time-dependent (intra-wave) equi-phase mean measurements of sand concentrations under sheet flow conditions in oscillatory motion. The experiments were carried out in a large oscillating water tunnel with a sand bed for three cases, including two cases in which asymmetrical oscillatory flow conditions were employed (C1 and C2). Flow conditions were similar for C1 and C2 in the sense that both cases involved regular-asymmetric 2nd order Stokes waves with almost the same degree of asymmetry (0.66 and 0.64 respectively) and almost the same root mean square value of the horizontal velocity (0.58 m/s and 0.60 m/s respectively).

Figures 10.7 and 10.8 present details of the comparison between the equi-phase sand concentration values determined experimentally and by computation together with the values of the horizontal velocity as measured during the experiment at an elevation of 20 cm above the initial undisturbed bed.

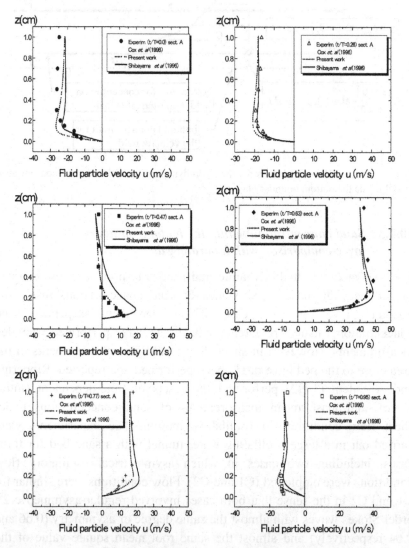

Figure 10.7 Comparison of numerical and laboratory results for the horizontal velocity inside BBL in the surf zone – Section A in the transition zone in the surf zone.

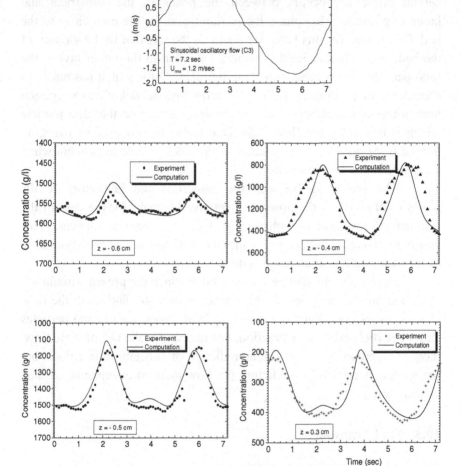

Figure 10.8 Outer velocity and temporal variation of sediment concentration inside BBL - case C3 (laboratory data from Ribberink and Al-Salem, 1992).

As can be observed, the numerical model acceptably predicts intra-wave sand concentrations. For the symmetrical flow, the two relatively symmetric sand concentration peaks could be predicted and unlike the case of asymmetrical flow, there is no observed difference between the accelerating and decelerating phase of the oscillatory motion. Still, for the case of the asymmetric oscillatory movement, a certain phase lag occurs between the peaks of the numerical and laboratory results. This phase lag is significant for the area close to the bed. One reason for this behavior could be the fact that in the vicinity of the bed, due to higher sand concentrations than in the upper layers, the sand particles might have some "inertial-delay" time which has not been considered in the present model. The term "inertial delay", as suggested here, refers to the effect of the inertial force acting on the sand particle which is driven by the flow field. Due to the presence of its mass, an inertial force produces a time-delay of the sand particle movement when compared to the fluid particle.

The phase lag between the numerical and laboratory results seems to diminish in the upward direction. However, for the case of the symmetric flow, such a phase lag does not seem to appear. This observation might lead one to the intuitive conclusion that the asymmetry of the flow might also induce a certain effect.

Here it should also be reminded that, since the present simulation is carried out for the case of a high mixture of water and sand, the flow characteristics are definitely influenced by the presence of sand particles and the particle-particle interaction. For the case of the asymmetric flow, there is a certain degree of underprediction of concentration values, but the model succeeded in predicting the flow-induced asymmetry of sand concentration.

10.3 Mud Transport

10.3.1 *Mud behavior in coastal area*

Bottom surfaces of coastal environments are often covered with certain layers of soft mud. In estuaries and river mouth areas of large rivers such as Yangtze river in China and Ganges river in Bangladesh, or in large bay areas, most of the coastal bottom surface is covered with soft mud. In these areas, when the surface water waves travel onto a bottom of soft mud, an inter-surface wave between the water layer and the mud layer is generated. In the interface, a fluid mud layer with high water content is formed under stormy conditions. The inter-surface wave causes mass transport in the mud layer. This type of mud mass transport as well as the suspended mud transport in the water layer are considered to be the main mechanisms that transport mud in coastal environment. Mud transport by the former (mass transport in mud layer) is found to be larger in quantity compared to the latter under many conditions of soft mud and therefore its consideration is important. The behavior of suspended mud is also important when we analyze coastal environment problems or the effect of construction works to the environment.

As stated in the previous part of this section, the mechanism of mud transport can be classified into two types: (1) mud mass transport in mud layer, and (2) suspended mud transport in water layer. In the following sections, emphasis will be given on the first transport type due to its greater significance to the transport processes.

But before we proceed with our discussion on mud behavior, we shall first discuss the characteristics of mud. Otsubo and Muraoka (1986) classified mud into two groups according to their characteristics of settling form, flow curve (shear rate - shear stress curve) and re-suspension behavior under uni-directional flow. The main factor to control these characteristics is the nature of the cation attached to the particle surface. The first group is represented by Kaolinite with Al^{3+} Ca^{2+} or H^+, and the second group by Bentonite which consists of Na^+ Montmorillonite. The first group forms an apparent mid-surface between the water layer and mud layer when mud particles settle down and it has yield value in flow curve. The second group does not form a mid-surface and does not have a yield value. Shibayama *et al.* (1986) suggested that both suspended load and mass transport in mud layer should be considered for the first group and suspended load alone should be considered for the second group. It is also suggested by Shibayama *et al.* (1986) that in the coastal environment, since the salinity of sea water supplies enough number of cations, most of mud behaves like the first group. Therefore if we consider mud behavior in the coastal environment, we mainly consider the first group.

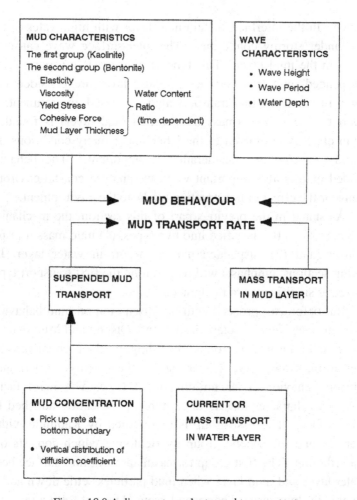

Figure 10.9 A diagram to evaluate mud transport rate.

Figure 10.9 shows the diagram of a model for mud transport due to waves. In the figure, it is stated that the mud behavior or mud transport rate is governed by the mud characteristics and wave conditions. In order to calculate the transport rate, the mass transport in the mud layer and suspended mass transport should be evaluated. The suspended transport rate can be evaluated by the information of mud concentration distribution and current or mass transport in the water layer.

10.3.2 *Cross-shore mud transport and beach deformation (Soltanpour, Shibayama and Noma, 2003)*

During the past decades, a number of studies have been conducted on different aspects of wave-mud interaction such as mud mass transport and wave attenuation. High wave attenuation at muddy coasts has been observed and the reason could be explained by interaction between water waves and soft mud at the seabed. Notable energy dissipation takes place within the mud bed (Gade, 1958; Tubman and Suhayda, 1976). On the other hand, it is also believed that waves are responsible for destructive mud slides which cause costly damage to offshore platforms (Sterling and Strohbeck, 1973). Studies on wave-mud interaction have been done and several models have been proposed based on the modeling of rheological property of soft mud such as the viscous fluid models by Gade (1958) and Dalrymple and Liu (1978), elastic models by Mallard and Dalrymple (1977) and Dawson (1978), a poro-elastic model by Yamamoto *et al.* (1983), visco-elastic models by Macpherson (1980), Hsiao and Shemdin (1980) and Maa and Mehta (1990), Bingham fluid models by Krone (1963), Tsuruya *et al.* (1987) and Mei and Liu (1987), and the visco-elastic-plastic model by Shibayama *et al.* (1990).

For rheological property measurement, Krone (1963), Tsuruya *et al.* (1986) and Otsubo and Muraoka (1988) conducted experiments with rotary viscometers and showed that mud behaves as a Bingham fluid. But the condition of rotary motion of mud in these experiments is different from the real condition on the natural coasts where water waves act as the driving forces on the mud bed. Shibayama and An (1993) used oscillating type viscometer to cope with real condition.

However, all of these studies have been conducted on horizontal beds. In order to approach the real field problems, where the fixed bed and mud surface are not necessarily horizontal, the influence of the bottom configuration on mud mass transport and wave height transformation should be examined. Although recently there have been a few studies on mud bed deformation (Shen, 1993), dynamics of mud shore profiles (Lee and Mehta, 1997), and downward flow of mud layers (Kessel and Kranenburg, 1996), there was no attempt to quantitatively examine reconfiguration of mud profiles.

Soltanpour *et al.* (2003) simulated the various features of wave-mud interaction on fine-grained shore profiles including wave height attenuation, wave-induced mud mass transport, gravity-driven flow of fluid mud and the reconfiguration of profile shape. A two-dimensional mud beach deformation model was presented considering the transport of fluid mud under continued wave action and downward gravity force. The rheological constitutive equations of visco-elastic-plastic model (Shibayama *et al.*, 1990) were selected for numerical simulation. Wave flume experiments were carried out and the results were utilized for the verification of the numerical model.

a) Rheological equations

Due to the complexity of mud behavior, different constitutive equations have been assumed for the prediction of the response of muddy beds. The large number of different proposed models themselves shows that a satisfactory rheological model has not been obtained yet and more studies on rheological behavior of mud are necessary. Here the visco-elastic-plastic model (Shibayama and An, 1993) is taken as an example in the numerical modeling of wave-mud interaction and gravity driven flow.

The constitutive equations of visco-elastic-plastic model are expressed as,

$$\sigma_{ij} = 2\mu_e \dot{e}_{ij} \tag{10.39}$$

$$\mu_e = \begin{cases} \mu_1 + \dfrac{iG}{w} & \left(\dfrac{1}{2}\sigma_{ij}\sigma_{ij} \leq \tau_y^2\right) \\[3mm] \mu_2 + \dfrac{\tau_y}{\sqrt{4|\Pi_e|}} & \left(\dfrac{1}{2}\sigma_{ij}\sigma_{ij} > \tau_y^2\right) \end{cases} \tag{10.40}$$

where μ_e is the apparent viscosity, σ_{ij} is the deviator part of stress tensor, $\dot{e}_{ij}$ is the strain tensor, G is the elastic modulus, w is the radian frequency, τ_y is the yield stress. μ_1 is the viscosity of mud in the viscoelastic state, μ_2 is the viscosity of mud in the viscoplastic state and $4|\Pi_e|$ is expressed as,

$$4|\Pi_e| = 2\left(\frac{\partial u}{\partial x}\right)^2 + 2\left(\frac{\partial w}{\partial z}\right)^2 + \left(\frac{\partial u}{\partial z} + \frac{\partial w}{\partial x}\right)^2. \tag{10.41}$$

The viscous modulus and the elasticity modulus were proposed by (Shibayama and An, 1993) as a function of water content of mobile mud and wave period.

b) Wave-mud interaction modeling

Following Tsuruya *et al.* (1987), the fluid system is divided into N layers in which the water layer is represented by $N = 1$ (see Fig. 10.10). The vertical distribution of mud properties can be considered using different mud characteristics in the horizontal sub-layers of the mud layer. The linearized Navier-Stokes equations, neglecting the convection term, and the continuity equation for an incompressive fluid can be expressed as

$$\frac{\partial u_j}{\partial t} = -\frac{1}{\rho_j}\frac{\partial p_j}{\partial x} + v_j\left(\frac{\partial^2 u_j}{\partial^2 x^2} + \frac{\partial^2 u_j}{\partial^2 z^2}\right) \tag{10.42}$$

$$\frac{\partial w_j}{\partial t} = -\frac{1}{\rho_j}\frac{\partial p_j}{\partial z} + v_j\left(\frac{\partial^2 w_j}{\partial^2 x^2} + \frac{\partial^2 w_j}{\partial^2 z^2}\right) \tag{10.43}$$

$$\frac{\partial u_j}{\partial x} + \frac{\partial w_j}{\partial z} = 0 \tag{10.44}$$

where x and z are the horizontal and vertical coordinates, u and w are the horizontal and vertical components of orbital velocity, the subscript j indicates the layers, and the parameters t, ρ, v and p represent the time, density, kinematic viscosity of mud and dynamic pressure respectively.

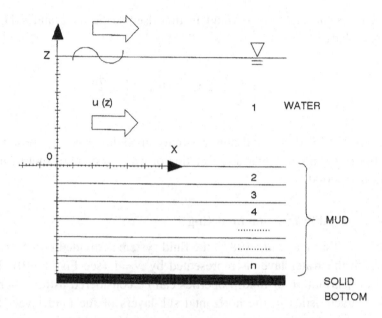

Figure 10.10 Definition sketch for mud transport model.

c) Wave-induced mud mass transport

The wave-driven mass transport velocity consists of two components, namely, the Stokes' drift and the mean Eulerian velocity. On an inclined bed, mud sediment may flow down the slope or, under significant wave action, be advected shoreward as a sediment mass under the influence of gravity force as well as pressure gradients. The effect of the gravity force is also important for the horizontal beds since a small slope in the mud surface, which may be generated after the wave action, strongly affects the mass transport.

d) Beach profile change

The conservation equation of sediment mass is employed to calculate beach profile change,

$$\frac{\partial h}{\partial t} = -\frac{\partial q_s}{\partial x} \qquad (10.45)$$

where h is the water depth and q_s is the rate of cross-shore volumetric sediment transport per unit length of shoreline.

In the present section, soft mud transport in coastal environment is explained as a brief review of the previous works. It is concluded that the mud mass transport can be evaluated by the visco-elastic-plastic model with the effect of the mud bottom slope. Figure 10.11 shows a schematic view of the numerical model to calculate the transport rate.

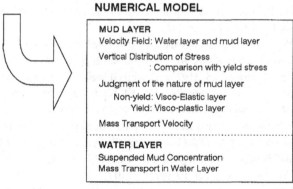

INPUT DATA

TOPOGRAPHY
Water Layer : Depth
Mud Layer : Vertical Distribution of Mud Density

MUD
Viscosity (Function of density and strain rate)
Elasticity (Function of density and wave period)
Yield Stress (Function of density)

WAVE AND CURRENT
Wave Height
Wave Period
Total Flow Discharge

NUMERICAL MODEL

MUD LAYER
Velocity Field: Water layer and mud layer
Vertical Distribution of Stress
: Comparison with yield stress
Judgment of the nature of mud layer
Non-yield: Visco-Elastic layer
Yield: Visco-plastic layer
Mass Transport Velocity

WATER LAYER
Suspended Mud Concentration
Mass Transport in Water Layer

OUTPUT DATA

TOTAL TRANSPORT RATE
Mud Mass Transport Rate in Mud Layer
Suspended Mud Transport in Water Layer

Figure 10.11 Schematic view of numerical model for mud transport.

References

Asano, T. (1990): Two-phase flow model on oscillatory sheet-flow, *Proc. of the 22nd Coastal Engineering Conference*, ASCE, pp. 2372-2384.

Coussot, P. (1994): Steady, laminar, flow of concentrated mud suspensions in open channel, *Jour. Hydra. Research*, Delft, 32(4), pp. 535-559.

Cox, D.T., Kobayashi, N. and Okayasu, A. (1996): Bottom shear stress in the surf zone, *J. of Geopysical Res.*, Vol. 101, No. C6, pp. 14337-14348.

Cristoferssen, J.B. and Jonsson, I.G. (1985): Bed friction and dissipation in combined current and wave motion, *Ocean Eng.*, Vol. 12, No. 5, pp. 387-423.

Dalrymple, R.A. and Liu, P.L.-f. (1978): Waves over muds, a two-layer fluid model, *Jour. Physical Oceanography*, Vol. 8, pp. 1121-1131.

Daubert, O., Haugeland, A. and Cahouet, J. (1982): Water waves calculation by Navier-Stokes calculation, *Proc. of the 18th Coastal Eng. Conference*, ASCE, pp. 832-845.

Davies, A.G. (1986): A model of oscillatory rough turbulent boundary flow, *Est. Coast. Shelf Sci.*, Vol. 23, pp. 353-374.

Dawson, T.H. (1978): Wave propagation over a deformable sea floor, *Ocean Engineering*, Vol. 5, 1978, pp. 227-234.

Duy, N.T. and Shibayama, T. (1997): A convection-diffusion model for suspended sediment in the surf zone, *J. of Geophys. Res.*, Vol. 102, No. C10, pp. 23169-23186.

Fredsøe, J. and Deigaard, R. (1992): *Mechanics of coastal sediment transport*, World Scientific Publishing Co. Pte. Ltd., 369 pp.

Gade, H.G. (1958): Effects of a non-rigid, impermeable bottom on plane surface waves in shallow water, *Jour. Marine Res.*, Vol. 16, No. 2, pp. 61-82.

Grant, W.D. and Madsen, O.S. (1979): Combined wave and current interaction with a rough bottom, *J. of Geophys. Res.*, Vol. 84, No. C4, pp. 1797-1808.

Hinze, J.O. (1975): *Turbulence*, McGraw-Hill, New York, p. 463.

Horikawa, K., Watanabe, A. and Katori, S. (1982): Sediment transport under sheet flow conditions, *Proc. of the 18th Coastal Eng. Conference*, ASCE, pp. 1335-1352.

Hsiao, S.V. and Shemdin, O.H. (1980): Interaction of ocean waves with a soft bottom, *Jour. Phys. Oceanogr.*, Vol. 10, pp. 605-610.

Jayaratne, M.P.R. and Shibayama, T. (2004): An Evaluation Method of Suspended Sediment Concentration in and outside the Surf Zone, *Proc. Coastal Eng. Conf.*, pp. 1715-1727.

Jonsson, I.G. (1964): Wave boundary layers and friction factors, *Proc. 10th Coastal Eng. Conf*, Vol. 1, pp. 127-148.

Justesen, P. (1991): A note on turbulence calculations in the wave boundary layer, *J. Hydr. Res.*, IAHR, Vol. 29(5), pp. 699-711.

Kajiura, K. (1968): *A model of the bottom boundary layer in water waves*, Bull. Earthquake Res. Inst., Univ. of Tokyo, Vol. 45, pp. 75-123.

Kamphuis, J.W. (1975): Friction factors under oscillatory waves, *J. of Waterways, Harbors Coastal Eng. Div. Am. Soc. Civil Eng.*, 101(WW2), pp. 135-144.

Kessel, T.V. and Kranenburg, C. (1996): Gravity current of fluid mud on sloping bed, *Jour. Hydraulic Engineering*, ASCE, Vol. 122, No. 12, pp. 710-717.

Kraus, N.C. and Smith, J.M. (1994): *SUPERTANK Laboratory Data Collection Project*, Vol. I, Coastal Engineering Research Center, USA, pp. 211-233.

Krone, R.B. (1965): *A study of rheological properties of estuarial sediments*, Tech. Bulletin No. 7, Committee on Tidal Hydraulics, U. S. Army Corps of Engineers, Waterways Experiment Station, Vickburg.

Larson, M. (1996): Closed form solution for turbulent wave boundary layer, *Proc. of the 25th Coastal Engineering Conference*, ASCE, pp. 3244-3256.

Lee, S.C. and Mehta, A.J. (1997). Problems of characterizing dynamics of mud shore profiles, *Journal of Hydraulic Engineering*, ASCE, Vol. 123, No. 4, pp. 351-361.

Maa, J.P.-Y. and Mehta, A.J. (1990): Soft mud response to water waves, *Jour. Waterway, Port, Coastal and Ocean Eng.*, ASCE, Vol. 116, No. 5, pp. 634-650.

Macpherson, H. (1980): The attenuation of water waves over a non-rigid bed, *Jour. Fluid Mech.*, Vol. 97, Part 4, pp. 721-742.

Madsen, O.S. and Grant, W.D. (1976): Quantitative description of sediment transport by waves, *Proc. 15th Coastal Eng. Conf.*, pp. 1093-1112.

Mallard, W.W. and Dalrymple, R.A. (1977): Water waves propagating over a deformable bottom, *Proc. 9th Annual Offshore Tech. Conf.*, Houston, Texas, pp. 141-146.

Mcpherson, H. (1980): The attenuation of waves over a non-rigid bed, *JFM*, 97-4: pp. 721-742.

Mei, C.C. and Liu, K.-F. (1987): A bingham-plastic model for a muddy seabed under long waves, *Jour. Geophys. Res.*, Vol. 92, No. C13, pp. 14,581-14,594.

Nadaoka, K. and Yagi, H. (1990): Single-phase fluid modelling of sheet flow toward the development of "numerical mobile bed", *Proc. of the 22nd International Conference of Coastal Engineering*, ASCE, pp. 2346-2359.

Nguyen, T. Duy and Shibayama, T. (1997): A Convection-Diffusion Model for Suspended Sediment in the Surf Zone, *JGR*, Ocean, 102(C10), 23169-23186.

Nielsen, P. (1988): Three simple model of sediment transport, *Coastal Eng.*, Elsevier Science, No. 12, pp. 43-62.

Otsubo, K. and Muraoka, K. (1985): Physical properties and critical shear stress of cohesive bottom sediments, *Proc. JSCE*, 363, pp. 225-234 (in Japanese).

Otsubo, K. and Muraoka, K. (1988). Critical shear stress of cohesive bottom sediments, *Jour. Hydraulics Eng.*, Vol. 114, No. 10, pp. 1241-1256.

Rattanapitikon, W. and Shibayama, T. (1998): Energy dissipation model for regular and irregular breaking waves, *Coastal Eng. in Japan*, JSCE, Vol. 40, No. 4, pp. 327-346.

Ribberink, J. and Al-Salem, A.A. (1992): *Time-dependent sediment transport phenomena in oscillatory boundary-layer flow under sheet flow conditions, Part IV*, Data Report H 840.20, DELFT Hydraulics, Netherlands.

Ross, M.A. and Mehta, A.J. (1990). *Fluidization of soft estuarine mud by waves. In: The Microstructure of Fine Grained Sediments: From Mud to Shale*, R.H. Bennett ed., Springer-Verlag, New York, pp. 185-191.

Sakakiyama, T. and Bijker, E.W. (1989). Mass transport velocity in mud layer due to progressive waves, *Proc. 21st Coastal Eng. Conf.*, ASCE, pp. 614-633.

Sato, S., Homma, K. and Shibayama, T. (1990): Laboratory study on sand suspension due to breaking waves, *Coastal Eng. in Japan*, JSCE, Vol. 33, No. 2, pp. 219-231.

Shen, D. (1993). *Study on mud mass transport and topography change of muddy bottom due to waves*, Ph.D. dissertation, University of Tokyo, 167 pp.

Shibayama, T. and Horikawa, K. (1980): Bed load measurement and prediction of two dimensional beach transformation, *Coastal Eng. in Japan*, Vol. 23, pp. 179-190.

Shibayama, T. and Nguyen, N. An (1993): A Visco-Elastic-Plastic Model for Wave-Mud interaction, *Coastal Engineering in Japan*, JSCE, 36(1), 67-89.

Shibayama, T. and Nistor, I. (1998): Modelling of Time-Dependent Sand Transport at the Bottom Boundary Layer in the Surf Zone, *Coastal Engineering Journal*, JSCE, 40(3), 241-263.

Shibayama, T., Duy, N.T., Okayasu, A. and Nistor, I. (1996): A simulation of velocity profile in the turbulent wave boundary layer in the surf zone, *Proc. Coastal Eng.*, JSCE, Vol. 43(1), pp. 446-451 (in Japanese).

Shibayama, T., Okuno, M. and Sato, S. (1990): Mud transport rate in mud layer due to wave action, *Proc. 22nd Coastal Eng. Conf.*, ASCE, pp. 3037-3048.

Shibayama, T., Sato, S., Asada, H. and Temmyo, T. (1989): Sediment Transport Rate in Wave-Current Coexistent Field, *Coastal Engineering in Japan*, JSCE, 32(2), 161-172.

Shibayama, T., Takikawa, H. and Horikawa, K. (1986): Mud mass transport due to waves, *Coastal Eng. in Japan*, 29, pp. 151-161.

Shibayama, T. and Rattanapitikon, W. (1993): Vertical distribution of suspended sediment concentration in and outside the surf zone, *Coastal Eng. in Japan*, JSCE, Vol. 36, No. 1, pp. 49-65.

Soltanpour, M. (1999): *Two-dimensional modeling of mud profile processes*, Ph.D. dissertation, Yokohama National University, 165 pp.

Soltanpour, M., Shibayama, T. and Noma, T. (2003): Cross-shore mud transport and beach deformation model, *Coastal Engineering Journal*, JSCE, Vol. 45, No. 3, pp. 363-386.

Sterling, G.H. and Strohbeck, E.E. (1973): The failure of the South Pass 70"B" platform in hurricane Camille, *Proc. 5th Conf. on Offshore Tech.*, Houston, Texas, pp. 719-730.

Throwbridge, J. (1983): *Wave-induced turbulent flow near a rough bed; implications of a time-varying eddy-viscosity*, Doctoral Dissertation, Woods Hole Oceanographic Institution, Massachusetts, 246 p.

Tsuruya, H., Nakano, S. and Takahama, J. (1987): Interaction between surface waves and a multi-layered mud bed, *Rep. Port and Harbor Res. Inst.*, Ministry of Transport, Japan, Vol. 26, No. 5, pp. 141-142.

Tsuruya, H. and Nakano, S. (1987): Interactive effects between surface waves and a muddy bottom, *Proc. of Coastal Sediments*, 1, pp. 50-62.

Tubman, M.W. and Suhayda, J.N. (1976): Wave action and bottom movements in fine sediments, *Proc. 15th Conf. on Coastal Eng.*, ASCE, 2, pp. 1168-1183.

Vongvisessomjai, S. (1986): Profiles of suspended sediment due to wave action, *J. Waterways, P., C., and Ocean Eng.*, ASCE, Vol. 112, No. 1, pp. 35-53.

Yamamoto, T., Takahashi, S. and Schuckman, B. (1983): Physical model of sea-seabed interactions, *Jour. Eng. Mech.*, ASCE, Vol. 109, No. 1, pp. 54-72.

Exercises

Problem 10.1 Derive the on-offshore sand transport rate formula based on the power model. The final expression should be the relationship between the Shields parameter and the non-dimensional volumetric sand transport rate.

Problem 10.2 A wave with a period of 10 seconds and a deep water wave height of 1 m, is coming to a beach which consists of sand (sand diameter: 0.7 mm).

(1) Determine the wave height at the point where the water depth is 2 m.
(2) Determine Jonsson's wave friction factor and the Shields parameter at the point of (1).

The beach slope is 1/200. In order to stop longshore sand transport, a groin is planned to be constructed. Evaluate the length of groin required to stop all longshore sand transport under the above conditions (Give estimated groin length). Discuss whether it is appropriate or not to design a groin under the concept "to stop all longshore transport" from the engineering point of view.

Problem 10.3 Calculate the Shields parameter associated with a 10 second wave in a water depth of 4 m and a wave height of 1.2 m propagating over a sandy bottom of diameter 2 mm.

Discuss briefly whether the sand starts to move or not by using the method of Madsen and Grant (1976).

Problem 10.4

(1) Calculate the longshore transport rate by using the formula of Komar and Inman (1970) for the following condition.

Wave period: 10 seconds
At the breaking point: Wave height: 2 m
Water depth: 3.5 m
Wave angle: 25 degrees

(2) Calculate the longshore transport rate by the using empirical constant of Savege (1962) for the same condition as (1). Discuss briefly the reason for the difference between the results of these two formulas.

Chapter 11

Beach Erosion and Beach Accretion

11.1 Cases in Developing Countries

A comparative study was performed for coastal processes in developing countries including Japan in the period between 1950 and 1980, Thailand, and Vietnam. The examples of coastal erosion problems were collected, classified and discussed in relation to the stages of economic development in these countries. The time history of occurrences of engineering problems in the coastal area was compared with the socioeconomic development stage of each country. It was found that coastal problems were closely related with industrialization and developmental stages.

11.1.1 *General trend – Japan model*

The developing countries in the Asian region were accomplishing rapid economic development for a long period before the short recession period which started in 1997. Industry, agriculture, and resort development were experimented a rapid advance. As a result, the Asian region faces a rapid change in the coastal environmental not experienced by the human race before. Also in African countries, for example in Tanzania, we can observe the same type of coastal erosion problems. The coastal erosion problems are related, through a coastal engineer's view point, to human activity. In this chapter it will be shown how this erosion is closely related to the process of industrialization and the modernization of the society. Figure 11.1 shows the process of coastal erosion based on Japanese experience during the period of 1960 to 1980. The coastal region in Thailand and Vietnam are good examples on how to apply the model based on Japanese experience.

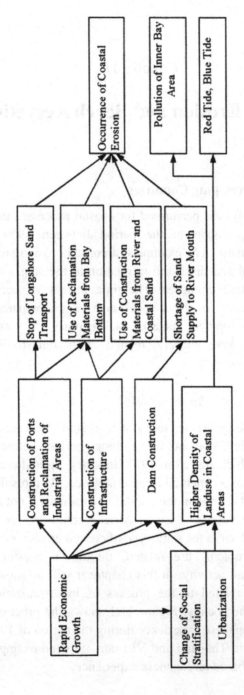

Figure 11.1 Industrialization process and coastal problems – Japan Model (Shibayama *et al.*, 1996).

The focus is placed on the coastal problems of Thailand, which experienced a big change during the 1980's (Shibayama *et al.*, 1996). Also, the case of Vietnam will be discussed in detail, which is similarly being subjected to significant changes as a result of the country's rapid economic development in recent years.

11.1.2 *The case of Thailand*

Historically, the first coastal problem was related to the maintenance of the waterway to the port of Bangkok and the maintenance of local small-scale fishing ports in the 1970's. Various coastal environment problems such as coastal erosion, waste water, the mangroves park construction and new industrial ports have occurred since 1985. It is possible to divide the source of the problems into two categories: (a) lack of environmental measures and (b) lack of technological knowledge. There are several cases from Japanese experiences of coastal problems which became good references for problem solving of the present situation of Thailand (Shibayama, 1995).

11.1.3 *The case of Vietnam (Shibayama and Tuan, 2001)*

The recent situation in Vietnam is comparable with that of Japan and Thailand. Coastal survey along the Vietnamese coast was performed in 1996 and compared with the report of 1992 and 1999. It appeared that as a result of rapid economic growth since 1990, the problems found are the same as those of Japan in the 1950's and of Thailand in the 1980's (refer to Table 10.1). An example is the coastal erosion occurred in Quang Ngai in December 2000.

Vietnam is experiencing both rapid economic and social changes since the promotion of Doi Moi (reform) policy, which started in 1986. Moreover, because the coast in Vietnam has been formed from the dynamic equilibrium of the erosion by waves and the deposit of sand supplied from large rivers and many small and medium rivers, a potential possibility of big coastal reaction to a slight change in land use exists. General views are presented regarding the following coastal problems for these last fifteen years.

1) The stage of 1992

In 1992, Quang and An reported the situation of the coast in Vietnam. At that time, securing the sea route for the Saigon river port, the blockage of a river mouth, and the coastal erosion in a limited part of the coast were some of the major coastal problems. The decrease in the mangroves by the influence of the chemical arms used by the US during the war and the deforestation in recent years advanced the coastal erosion though the lagoon lake and the mangrove woods. Erosion problems were still not observed along the coastline. At this stage, there were worries that damage to the coastline, which the Philippines and Thailand experienced in the past though the construction of shrimps pond, might have also started for the same reason in Vietnam. At this time when the ponds were constructed, no standard for natural environment protection existed.

At the stage of 1992, major investment was being planned for Hai Phong port in the north and Vung Tau port in the south. Also the maintenance of the transportation route in the Mekong River was important, particularly in the mouth of this river. It would have been important to observe the coastline stability carefully because it appears likely that in the future therewould be a more dense utilization of the coastal resources.

The appearance of concrete signs of severe coastal problems was not reported at by Quang and An (1992). Hence at this stage it appears that the problems which have now been recorded along the Vietnamese coast had not appeared up to this date.

2) The stage of 1996

However, in 1996 Shibayama *el al.* conducted a major survey of the Vietnamese coast from Ho Chi Minh City to Hanoi. The following section outlines the conclusions of this survey starting from the south of the country and moving northwards.

Intrusion of salt water from the Mekong river mouth is creating a threat to agriculture development in the Mekong delta region, which is regarded as one of the major rice production areas in Vietnam. Furthermore, the Mekong River, as an international river, is an important

Table 11.1 Time histories of three countries.

Years	Japan	Thailand	Vietnam
1870	Waterway maintenance in Niigata Port		
1880			
1890			
1900			
1910			
1920	Beach Erosion in Niigata Coast		
1930			
1940			
1950	59 Storm Surge by Ise Bay Typhoon Protection against Storm Surge		
1960	Artificially Excavated Port 63 Kashima Port Reclamation Work in Tokyo Bay	65 Maintenance of River Mouth of Chao Phraya	
1970	71 Detached Breakwater in Kaike Measures for Beach Protection	Maintenance of River Mouth	
1980		88 Laem Chabang Port Construction of Deepsea Port	
1990		90 Survey of Coastal Erosion Appearance of Coastal Erosion 90 Erosion in Phuket Island	94 Survey of Mekong River Mouth
2000			99 Beach Erosion in Lang Co

traffic route. The Mekong River splits into nine branches as it reaches the sea; however, the main routes from sea to the inland area are through Cua Tieu and Cua Dinh An. Due to the sand deposit and longshore sand transport in the south-west direction, the maintenance of this route is difficult.

A huge dune exits in the coastline in Mui Ne which is eastward from Ho Chi Minh City. The coastline in the vicinity of Mui Ne has sufferred coastal erosion in recent years, and the palm trees in the coast have collapsed due to this erosion.

The supply of sand is steady in general, and the shape of the coast is steady from the Vietnamese central part to the northern part though coastal protection facilities have not yet been installed.

A lagoon lake has been formed by the development of spits in Tuy An. Moreover, the development of a sand spit can be observed in Sa Huynh, Lang Co, and Ky Anh. Since the development of sand spits has also been observed and it appears that the central coastline in Vietnam is formed by the sand supply from the rivers. Therefore, it is forecasted that the decrease of sand and sediment supply from river due to the construction of dams for hydropower generation, flood protection measures and the exploitation of sands as a construction materials would lead to coastal erosion problem (refer to Fig. 11.1). This problem has already been observed in the areas where the industrial development has been more remarkable. An example is the case of the coast around Mui Ne.

In 1999, an example of construction problems concerning coastal protection was reported. The barrier in the lagoon lake of Tuy An collapsed due to the flood in November 1999, and the restoration of the coast was necessary.

11.1.4 *Discussions*

The main point of this chapter is how "the appearance of coastal erosion problems is different according to the developmental stage of the modernization and it can be clarified by making comparative study of the situation in countries at different stages of industrialization process". The historical change of GDP of Vietnam, Thailand and several other Asian countries from 1970 to 2001 was analyzed. The GDP growth figure indicates the rapid change of social structure in the middle 1990's for the case of Vietnam.

Figure 11.1 relates the coastal environment problems with the activities of economic development. The drawing of this figure is based on the experience of Japan since 1945. The process shown in it was a working hypothesis and it was examined for different countries. Assuming that the development in each country goes through a common process, the time of appearance of each coastal problem is different from country to country due to the developmental stage that each country finds itself in. In other words, it is thought that coastal environmental problems can be displayed as a time series by making the economic developmental stage as the time coordinates. By analyzing the process, it will be possible to stop or modify the process and get better results by using different developmental policies based on the Japanese experience.

Table 11.1 shows the time history of coastal problems in Japan, Thailand and Vietnam. From the table, the time difference from Japan to Thailand is around 20 – 40 years and from Thailand to Vietnam around 10 years. In Thailand or Vietnam, large scale development projects are conducted one after another without considering long-term negative effects on environment. Japan has been experienced a big change of environmental system, which could not be predicted at the time because the coastal region of Japan has been positively and gradually developed since the Meiji era started from 1868. In particular, a rapid change in the coast environment, which Japan has been experiencing since the rapid economic growth period after the World War II, is a useful lesson to solve the coastal environment problems which Asian nations are facing now. Learning from the experience of Japan can help coastal engineers in other Asian countries to prevent coastal problems.

11.2 Sediment Production in River Basin and River Sediment Transport to Coasts (Nobert and Shibayama, 2006)

In this section, a method to estimate total sediment discharge from a river basin to the coastal area is described. The model consists of soil erosion estimation and transport mechanism components, and a one-dimensional river profile change component. The input parameters for the model are derived from rainfall data, land use/land cover data, soil data and elevation data.

11.2.1 *Introduction*

Watershed sediment yield is given by surface erosion rates and is also the primary source of sediment discharge to the river mouths. The hydrologic processes of rainfall and runoff drive the surface erosion process. Surface erosion by water embodies the process of detachment, transportation, and deposition of soil particles by the erosive and transport agents of raindrop impact and runoff over soil surface.

11.2.2 *Methodology*

A case study of the Sakawa river basin in Japan is used to explain the methodology. Using the digital elevation data for the river basin with spatial resolution of 50mx50m, the basin is divided into 19 sub-basins as shown in Fig. 11.2 using ArcView GIS software. A soil erosion calculation was then performed for each subbasin. The area of the basin is 571.6 km^2 and the river originates from Fuji, Tanzawa and Hakone mountains.

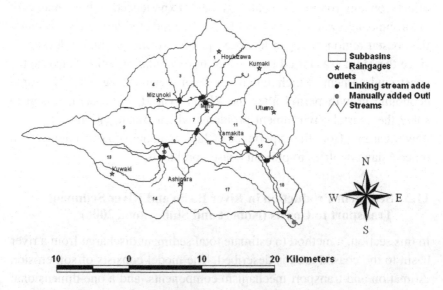

Figure 11.2 Subbasins in Sakawa river basin delineated from Digital Elevation Data.

(a) Soil Erosion equations

In this method the Modified Universal Soil Loss Equation (MUSLE), (Williams, 1975, 1985) is used to estimate soil erosion caused by rainfall and runoff

$$sed = 11.8(Q_{surf} \times q_{peak} \times area_{hru})^{0.56} \times K_{USLE} \times C_{USLE} \times P_{USLE} \times LS_{USLE} \quad (11.1)$$

where *sed* is the sediment yield on a given day (tons), Q_{surf} is the surface runoff volume (mm/ha), q_{peak} is the peak runoff rate (m^3/s), K_{USLE} is the soil erodibility factor (ton $\cdot$ m^2 $\cdot$ hr/(m^3 $\cdot$ ton $\cdot$ cm), $area_{hru}$ is the area of the sub-basin (ha), LS_{USLE} is the slope length factor(m) and C_{USLE} and P_{USLE} are land cover and land management factors respectively.

(b) Runoff volume (Q_{surf})

The runoff volume (Q_{surf}) is estimated using the Soil Conservation Service (SCS) curve number equation (SCS, 1972)

$$Q_{surf} = \frac{(R_{day} - I_a)^2}{(R_{day} - I_a + S)} \quad (11.2)$$

where I_a is the initial abstraction which includes surface storage, interception and infiltration prior to runoff (mm), R_{day} is the daily rainfall and S is the retention parameter.

The retention parameter varies spatially due to changes in soils, land use, management and slope and temporarily due to changes in soil water content. The retention parameter is defined as:

$$S = 25.4 \left(\frac{1000}{CN} - 10 \right) \quad (11.3)$$

where *CN* is the curve number and it represents the potential for storm water runoff in the drainage area.

(c) Peak runoff rate (q_{peak})

The peak runoff rate is calculated with a modified rational equation (USDA-SCS 1986):

$$q_{peak} = \frac{\alpha \times q \times A}{3.6 t_c} \tag{11.4}$$

where q_{peak} is the peak runoff rate (m^3/s); q is runoff (mm); A is sub-basin area (km^2), t_c is time to concentration (hr) and α is the dimensionless parameter that expresses the proportion of total rainfall that occurs during t_c. The time of concentration can be calculated as a sum of the overland flow time (the longest time needed for the overland flow to reach the channel) and the channel flow time (the longest time it takes for water to travel from the upland channels to the outlet). The overland time of concentration can be calculated as (Arnold *et al.*, 1995):

$$t_{ov} = \frac{L^{0.6} \times n^{0.6}}{18 \times S^{0.3}} \tag{11.5}$$

where t_{ov} is the overland time of concentration (hr), L is the slope length (m), n is the Manning's roughness coefficient for overland flow, S is the slope steepness (m/m). The channel time of concentration is calculated as:

$$t_{ch} = \frac{0.62 \times L \times n^{0.75}}{A^{0.125} \times S_{ch}^{0.375}} \tag{11.6}$$

where t_{ch} is the channel time of concentration (hr), L is the channel length from the most distant point to the watershed outlet (km), n is the Manning's roughness coefficient, A is the watershed area (km^2), S_{ch} is the channel slope (m/m).

(d) Soil erodibility: K_{USLE} factor

The method used for estimating the soil erodibility factor is that used in the EPIC model (Sharpley and Williams, 1990).

$$K = \frac{1}{7.6}\left\{0.2+0.3\exp\left[-0.0256SAN\left(1-\frac{SIL}{100}\right)\right]\right\}\left(\frac{SIL}{CLA+SIL}\right)^{0.3}$$

$$\times\left(1.0-\frac{0.25OM}{orgC+\exp(3.72-2.95OM)}\right)$$

$$\times\left(1.0-\frac{0.7SN}{SN+\exp(-5.51+22.9SN)}\right) \quad\quad (11.7)$$

where $SN = 1.0 - SAN/100$ and SAN, SIL, CLA and OM are the percentage content of sand, silt, clay and organic matter respectively.

(e) The Slope length and steepness factor: LS factor

This factor is calculated using the following equation:

$$LS = \left(\frac{L}{22.1}\right)^m (65.41\sin^2\theta+4.56\sin\theta+0.065) \quad\quad (11.8)$$

where L is the slope length (in meters), m is the exponential term, and θ is the angle of the slope. The exponential term, m, is calculated with the following equation:

$$m = \left(\frac{F}{1+F}\right) \text{ and } F = \left(\frac{\sin\theta/0.0896}{3(\sin\theta)^{0.8}+0.56}\right) \quad\quad (11.9)$$

11.2.3 *River profile change model*

(a) Flow equation

The one-dimensional steady flow, momentum equation is used for the flow calculation:

$$\frac{\partial H}{\partial x}+\frac{\partial}{\partial x}\left(\frac{\psi Q^2}{2gA^2}\right)+i_e = 0 \qu\quad (11.10)$$

$$i_e = \frac{u^2 n^2}{R^{4/3}} \qquad (11.11)$$

where: x is the distance from the river mouth, H is the water surface elevation (m), Q is the river discharge (m³/s), A is the cross-section area (m²), R is the hydraulic radius, i_e is the energy slope and ψ is the energy correction coefficient. Substituting Eq. (11.11) into equation (11.10), assuming $\psi \approx 1$ and $R \approx h$, Eq. (11.10) can be re-written as:

$$\frac{\partial H}{\partial x} + \frac{Q^2}{2g} \frac{\partial}{\partial x}\left(\frac{1}{B^2 h^2}\right) + \frac{Q^2 n^2}{B^2 h^{10/3}} = 0 \qquad (11.12)$$

$$H = \eta + h \qquad (11.13)$$

where η is the river is bed elevation (m) and h is the water depth (m).

(b) Sediment equations

(i) Bed load equations

The bed load transport rate per unit width is calculated by the Ashida and Michiue's formula (1972):

$$\frac{q_{Bi}}{\sqrt{sgd_i^3}} = p_i 17 \gamma_{*i}'^{3/2}\left(1 - \frac{\gamma_{*ci}}{\gamma_{*i}}\right)\left(1 - \frac{u_{*ci}}{u_*}\right) \qquad (11.14)$$

$$\frac{u_{*ci}^2}{u_{*cm}^2} = \left[\frac{\log 23}{\log\left(21\dfrac{d_i}{d_m} + 2\right)}\right] \frac{d_i}{d_m} \qquad (11.15)$$

where: q_{Bi} is the bed load (m³/s) per unit width, d_i the diameter of the bed material (m), p_i is the volumetric fraction of the sediment particles, s is the specific gravity of the sediment particles, γ_{*i} is the non-dimensional shear stress, γ_{*ci} the non-dimensional critical shear stress, u_{*ci} and the critical shear velocity.

(ii) Suspended load equations

The pick-up rate of the suspended load per unit area is calculated by the formula of Itakura and Kishi (1980).

$$q_{sui} = p_i K \left(\alpha_* \frac{\rho_s - \rho}{\rho_s} \frac{g d_i}{u'_*} \Omega_i - w_{fi} \right) \tag{11.16}$$

$$\Omega_i = \frac{\gamma'_{*i}}{B_{*i}} \frac{\int\limits_{a'}^{\infty} \xi \frac{1}{\sqrt{\pi}} \exp(-\xi^2) d\xi}{\int\limits_{a'}^{\infty} \frac{1}{\sqrt{\pi}} \exp(-\xi^2) d\xi} + \frac{\gamma'_{*i}}{B_{*i} \eta_0} - 1 \tag{11.17}$$

where $a' = B_{*i}/\gamma'_{*i} - 1/\eta_0$, $\eta_0 = 0.5$, $\alpha_* = 0.14$, $K = 0.008$, and

$$B_{*i} = \xi_i B_{*0}$$

$$\xi_i = \frac{\gamma_{*ci}}{\gamma_{*ci0}}$$

where, q_{sui} is the suspended volume from the bottom per unit area and w_{fi} is the fall velocity of suspended sediment according to diameter. The volumetric fraction of the bed material grain size is obtained from:

$$\delta \frac{\partial p_i}{\partial t} + p_i^* \frac{\partial \eta}{\partial t} + \frac{1}{1-\lambda} \left[\frac{1}{B} \frac{\partial (q_{Bi} B)}{\partial x} + q_{sui} - w_{fi} c_{bi} \right] = 0 \tag{11.18}$$

$$p_i^* = p_i; \ \partial \eta / \partial t \geq 0$$

$$p_i^* = p_{i0}; \ \partial \eta / \partial t < 0$$

where: δ is the thickness of the exchange layer and λ is the porosity of the bed material/void ratio, η is the river bed elevation (m). The time-dependent bottom profile change is obtained from the continuity of bed material transport:

$$\frac{\partial \eta}{\partial t} + \frac{1}{1-\lambda} \left[\frac{1}{B} \frac{\partial \sum_i (q_{Bi} B)}{\partial x} + \sum_i (q_{sui} - w_{fi} c_{bi}) \right] = 0 \tag{11.19}$$

11.2.4   *Sediment transport*

Sediment transport in the channel is a function of two processes, deposition and degradation, operating simultaneously in the reach. Sediment yield concentration from the watershed reaching the subbasin stream at the beginning of the time step is compared with the transport capacity (T_c) of that stream segment. If the initial concentration of sediments reaching the stream segment is greater than the transport capacity of the stream segment, deposition is the dominant process, otherwise degradation of the stream is the dominant process. At the end of the time step the river profile is updated and the process continues again. From the experimental analysis of the river transport capacity, Bagnold, 1977 established that the transport capacity of the river is a function of the peak velocity in the channel; based on the work by Bagnold, the transport capacity (T_c) for this case is calculated as (Arnold, et al., 1995),

$$T_c = aV_{pk}{}^b \qquad (11.20)$$

$$V_{pk} = \frac{q_{pk}}{A_{ch}} \qquad (11.21)$$

where A_{ch} is the cross-sectional area of flow in the channel (m^2) and q_{pk} is the peak flow rate in the channel (m^3/s), where T_c is the transport capacity (ton/m^3); V_{pk} is the peak channel flow velocity (m/s); a and b are the channel coefficients. The net amount of sediment deposited and eroded is calculated by,

$$sed_{dep} = (C_{in} - T) \times V_{ch} \qquad (11.22)$$

$$sed_{\deg} = (T - C_{in}) \times V_{ch} \times K_{ch} \times C_{ch} \qquad (11.23)$$

where sed_{dep} is the amount of sediment deposited in the reach segment (metric tons) and V_{ch} is the volume of water in the reach segment (m^3), $sed_{\deg}$ is the amount re-entrained in the reach segment (metric tons), K_{ch} and C_{ch} are the channel coefficients. Once the amount of deposition and degradation has been calculated, the final amount of sediment in the reach is determined:

$$sed_{ch} = sed_{ch,i} - sed_{dep} + sed_{deg} \qquad (11.24)$$

where sed_{ch} is the amount of suspended sediment in the reach (metric tons), $sed_{ch,i}$ is the amount of suspended sediment in the reach at the beginning of time period (metric tons). The amount of sediment transported out of the reach is calculated using

$$sed_{out} = sed_{ch} \times \frac{V_{out}}{V_{ch}} \qquad (11.25)$$

where sed_{out} is the amount of sediment transported out of the reach (metric tons), V_{out} is the volume of outflow during the time step (m^3), and V_{ch} is the volume of water in the reach segment (m^3). Figure 11.3 shows the flow-chart of the integrated model. All components are integrated to calculate total sand discharge produced in the river basin.

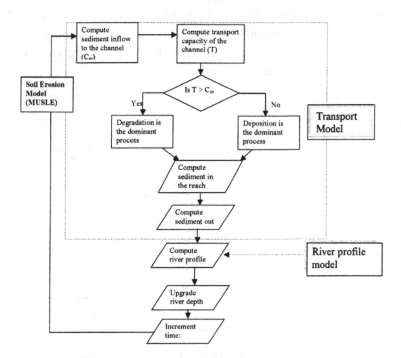

Figure 11.3 Flow chart for the integrated model.

11.2.5   *Sediment evaluation*

(1) Total sediment discharge to coastal environment

Generally, there is a strong relationship between sediment discharge and the river flow. The periods with high sediment discharge corresponds to high river flow and the periods with low sediments discharge correspond to low river flow periods. The scatter plot between the observed flow and the sediment discharge at the river mouth is shown in Fig. 11.4, and the correlation coefficient (R^2) between the two is 0.67. Figure 11.5 shows measured sediments data at the river mouth; annual sediment inflow from the river basin to the river mouth is calculated as a total of the deposited sediments and dredged sediments for each year. The comparison between the simulated sediments and the measured sediments at the river mouth for the period between 1990 and 2000 is shown in Fig. 11.6. Since the only available measured data is the yearly data, the comparison was done only on an annual basis. It can be observed that there is good agreement between the observed and measured sediment data. For the sediment transport mechanism used in this chapter, it was impossible to separate between the washload and the bed load. However, the total sediments discharge to the river mouth includes both washload and bed load.

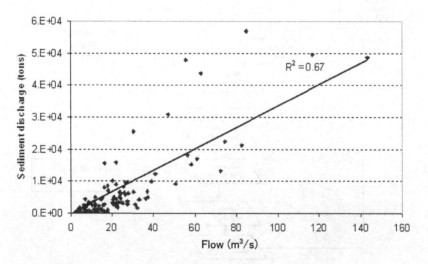

Figure 11.4 Scatter plot between stream flow and the sediment discharge.

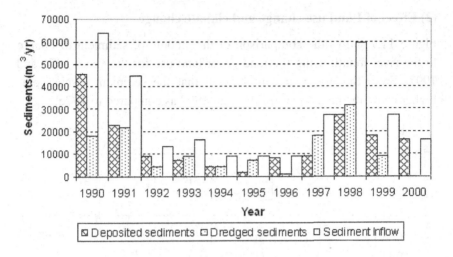

Figure 11.5 Measured sediments data at the river mouth.

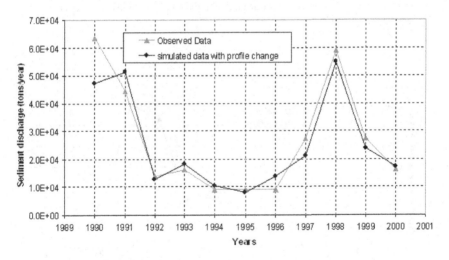

Figure 11.6 Observed and simulated sediments for the period 1990-2000 at Iizumi.

(2) Effect of Land use change and climate change

Figure 11.7 shows the comparison of the amount of sediment discharge to the river mouth using the land cover/land use for the years 1976 and 1997. From the graph, it can be seen that the sediment discharge is higher using the land cover for the year 1997 as compared to 1976.

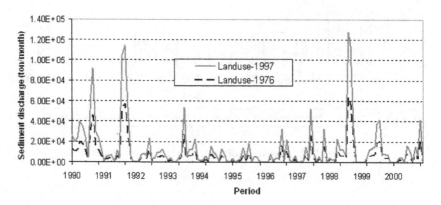

Figure 11.7 Comparison of sediment discharge at the river mouth using the land cover for the year 1976 and 1997.

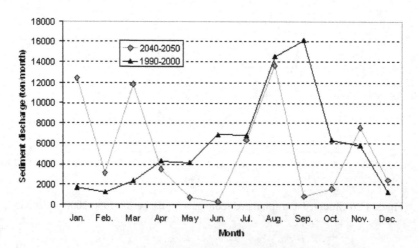

Figure 11.8 Simulation results for the years 1990-2000 and 2040-2050.

Figure 11.8 shows the total annual sediment discharge for the periods 1990-2000 and 2040-2050 for the present and future climate scenarios. From the figure, it can be seen that there is no clear trend for the sediment discharge at the river mouth that can be deduced despite the fact that there are some seasonal differences of sediment discharge due to changes in the rainy season.

References

Arnold, J.G., Williams, J.R. and Maidment, D.R. (1995): Continuous water and sediment routing model for large basins, *Journal of Hydraulic Engineering*, ASCE, 121(2), 170-183.

Bagnold, R.A. (1977): Bed Load transport by natural rivers, *Water Resources Research*, Vol. 13(2), 303-312.

Foster, G.R. (1982): Modelling the erosion processes. In: C.T. Haan (Editor), Hydrologic Modelling of Small Watersheds. ASAE Monograph, pp. 297-380. In Present and prospective technology for predicting sediment yield and sources: *Proceedings of the sediment yield workshop*, USDA Sedimentation Lab., Oxford, MS, November 28-30, 1972. ARS-S-40. p. 244-252.

Govindaraju, R.S. and Kavvas, M.L. (1992): Characterization of the rill geometry over straight hillslopes through spatial scales, *J. Hydro.*, Amsterdam, 130, 339-365.

Itakura, T. and Kishi, T. (1980): Open channel flow with suspended sediments, *Proc. of ASCE*, HY8, pp. 1325-1343.

Lu, H., Gallant, J., Prosser, I.P., Moran, C. and Priestley, G. (2001): *Prediction of Sheet and Rill Erosion over the Australian Continent, Incorporating Monthly Soil Loss Distribution*, Technical Report 13/01, CSIRO Land and Water, Canberra, Australia.

Moore, I.D. and Burch, G.J. (1986): *Modeling erosion and deposition: Topographic effects*, American society of agricultural engineers 000-2351:1624-1640.

Neitsch, S.L. *et al.* (2002): *Soil and Assessment Tool (SWAT) theoretical documentation*, version 2000, 506 pp.

Nicks, A.D. (1974): Stochastic generation of the occurrence, pattern, and location of maximum amount of daily rainfall, *Proc. Symp. Statistical Hydrology*, Aug.-Sept. 1971, Tucson, AZ, U.S. Department of Agriculture, Misc. Publ. No. 1275, pp. 154-171.

Nobert, J. and Shibayama, T. (2007): Integrated Model for Estimating Sediment Discharge to Coastal Area from River Basin, *Journal of Global Environmental Engineering*, 12, pp. 13-32.

Renard, K.G. *et al.* (1997): *Predicting soil erosion by water: A guide to conservation planning with the Revised Universal Soil Loss Equation (RUSLE)*, US Department of Agriculture, Agriculture Handbook No. 703, 404 pp.

Sharpley, A.N. and Williams, J.R. (1990): *EPIC-Erosion/Productivity Impact Calculator: USDA Technical Bulletin 1768*: 235 pp.

SCS (1972): *Soil Conservation Service (SCS)*, National Engineering Handbook, Section 4: Hydrology, US Department of Agriculture, USA.

Shibayama, T. and Le Trung Tuan (2001): A Comparative study of coastal processes in Asian countries, *Proc. of APAC (Asian and Pacific Coastal Engineering)*, pp. 909-917.

Shibayama, T., Shibayama, M. and Toue, T. (1996): Coastal problem appearance related to developmental stage of developing countries, *Proc. of Coastal Eng.*, JSCE, 43(2), 1291-1295 (in Japanese).

Shibayama, T. (1995): Three examples of Japanese experience of coastal environmental change due to construction work, *Proc. 4th COPEDEC*, pp. 2147-2152.

Shibayama, T. (Ed.) (1992): *Coastal processes in Asian region*, 174 p., Dept. of Civil Engineering, Yokohama National University, ISBN 4-9446476-00-8.

Shibayama, T. (2003): A comparative study of coastal problems in developing countries, *Proc. of COPEDEC VI*, Colombo, Sri Lanka.

Tayfur, G. and Singh, V.P. (2004): Numerical model for sediment transport over nonplanar, nonhomogeneous surfaces, *Journal of Hydrologic Engineering*, ASCE, Vol. 9, No. 1 (2004) 35-41.

Tran Minh Quang and Nguyen Ngoc An (1992): *The recent state of coastal development and coastal environmental problems in Vietnam*, pp. 53-62, Coastal Processes in Asian Region (Shibayama, T., Ed.).

Tuan, L.T. and Shibayama, T. (2003): Application of GIS to evaluate long-term variation of sediment discharge to coastal environment, *Coastal Engineering Journal*, Vol. 45, No. 2 (2003) 275-293.

Williams, J.R. (1975): *Sediment-yield prediction with universal equation using runoff energy factor*.

Woolhiser, D.A., Smith, R.E. and Goodrich, D.C. (1990): *KINEROS, A kinematic runoff and erosion model: Documentation and user manual*, ASR-77, U.S. Department of Agriculture, Washington, D.C., 130 pp.

Yang, D., Kanae, S., Oki, T., Koike, T. and Musiake, K. (2003): *Global potential soil erosion with reference to land use and climate changes*, Wiley Interscience, Hydrological Process. 17, 2913-2928 (2003).

Chapter 12

Natural Disasters in Coastal Environment

Coastal environments are vulnerable to three possible natural disasters, namely, tsunami, storm surge, and high wave attack due to storms. Since a storm surge is also caused by a Typhoon, a Hurricane or a Cyclone, there is a considerable probability that the second and the third types of disaster can arrive to the same area at the same time.

In the history of human race, many people have lost their lives, houses, properties, etc. due to coastal disasters. For example, in Nagoya city, Japan, 4,697 people were killed in 1957 due to the effects of a storm surge. In Netherlands, a similar disaster occurred in 1953. Recently, in 1991, a strong storm surge came to Bangladesh and in 2005 Hurricane Katrina came to the U.S. south coast. Hence, the Japanese experience became a strong motivation for the Japanese government to construct storm surge barriers along coastlines of concentrated urban coastal areas. Nonetheless, designing structures towards the protection against storm surges is a very important research topic in Coastal Engineering.

Tsunamis are caused by earthquakes in the continental shelf or coastal zones. In Japanese history, we can note the considerable number of tsunami disasters which the country has suffered. Between 1933 and 1983, there were several big tsunamis caused by earthquakes in the vicinity of the Japanese islands. Also, in recent times there has been the Indian Ocean tsunami, which occurred in December, 2004. The details of disaster survey of this tsunami will be described in Section 12.3. In 1960, there was a big earthquake in the offshore area of Chili, South America. The generated tsunami traveled through the pacific ocean via Hawaiian islands and finally struck the Japanese northern coast. Tsunami forecasting and protection, therefore, is also a major engineering problem.

12.1 Storm Surge

12.1.1 *Basic mechanism*

There are two basic mechanisms that can lead to the creation of a high tide under a storm: (1) pressure difference under low atmospheric pressure and (2) wind shear on the water surface.

The atmospheric pressure difference under a low pressure area $(p_\infty - p)$ is balanced by a pressure set-up of water η_{ps}.

$$\eta_{ps} = \frac{1}{\rho g}(p_\infty - p)$$

$$= 0.991(p_\infty - p) \tag{12.1}$$

where η_{ps}: pressure set-up, p_∞: normal pressure (hp), p: pressure at the site (hp).

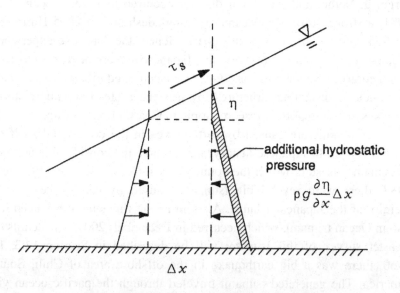

Figure 12.1 Wind set-up (vertically exaggerated).

The wind shear τ_s on the water surface is balanced by wind set-up (see Fig. 12.1 for definitions)

$$\rho g(h+\eta)\frac{\partial \eta}{\partial x}\Delta x = \tau_s \Delta x \qquad (12.2)$$

where ρ is water density and η is wind set-up.

With the assumption of $\eta \ll h$, wind set-up is evaluated by

$$\frac{\partial \eta}{\partial x} = \frac{\tau_s}{\rho g h}. \qquad (12.3)$$

12.1.2 Numerical simulation

For the storm surge calculation, a typhoon model should be set first. This includes the pressure distribution under the typhoon and route that the typhoon will take. Based on the model, the time history of wind distribution and pressure distribution can be calculated. Using this environmental information, the flow pattern is calculated by using the following equation set.

The mass conservation equation is (under long wave assumption)

$$\frac{\partial \eta}{\partial t} + \frac{\partial[(h+\eta)u]}{\partial x} + \frac{\partial[(h+\eta)v]}{\partial y} = 0. \qquad (12.4)$$

The momentum conservation equation in x-direction is

$$\frac{\partial u}{\partial t} + u\frac{\partial u}{\partial x} + v\frac{\partial u}{\partial y} - fv + g\frac{\partial \eta}{\partial x} + \frac{1}{\rho}\frac{\partial p}{\partial x} + \frac{\tau_x^b - \tau_x^s}{\rho(\eta+h)} - A_h\left(\frac{\partial^2 u}{\partial x^2} + \frac{\partial^2 u}{\partial y^2}\right) = 0$$

$$(12.5)$$

where f: corioli's coefficient, τ_x^b: bottom shear stress, τ_x^s: surface wind shear, A_h: horizontal mixing coefficient. For the y-direction, we have one more momentum conservation equation.

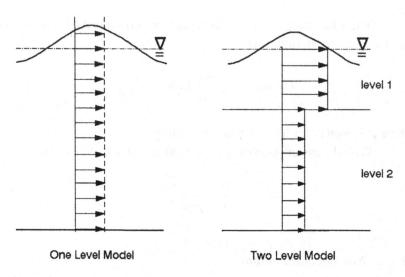

One Level Model   Two Level Model

Figure 12.2 The difference between one level model and two level model.

By using the above equation set, we can calculate the storm surge propagation by, for example, a finite difference scheme. Since the wind shear exerted on the surface causes a shear water flow, it is better to consider a three dimensional structure for the storm surge calculation.

Figure 12.2 shows the difference between a one-level model and a two-level model. In a two-level model, the water layer is divided into two parts and the vertical distribution of velocity is considered. The momentum conservation equation (12.5) is applied separately to the upper and lower layer. The shear stress between the two layers and the momentum exchange are also included in the two level calculations.

12.2   Tsunami

Tsunamis are generated by submarine earthquakes or submarine landslides. The displacement of the ocean bottom due to these two types of tsunami source is given as the initial condition of tsunami propagation. Since there is no wind shear, a one level model is enough to stimulate tsunami behaviour.

Mass conservation equation is given by Eq. (12.6) and momentum conservation equation for x-direction is given by Eq. (12.7) and for the y-direction by Eq. (12.8), but eliminating wind shear term and pressure term. The basic equations are thus the following;

Mass Conservation

$$\frac{\partial \eta}{\partial t} + \frac{\partial M}{\partial x} + \frac{\partial N}{\partial y} = 0 \qquad (12.6)$$

X-Momentum Flux Conservation

$$\frac{\partial M}{\partial t} + \frac{\partial}{\partial x}\left(\frac{M^2}{D}\right)\frac{\partial}{\partial y}\left(\frac{MN}{D}\right) + gD\frac{\partial \eta}{\partial x} + \frac{gn^2}{h^{7/3}}M\sqrt{M^2 + N^2} = 0 \quad (12.7)$$

Y-Momentum Flux Conservation

$$\frac{\partial N}{\partial t} + \frac{\partial}{\partial x}\left(\frac{MN}{D}\right)\frac{\partial}{\partial y}\left(\frac{N^2}{D}\right) + gD\frac{\partial \eta}{\partial x} + \frac{gn^2}{h^{7/3}}N\sqrt{M^2 + N^2} = 0 \qquad (12.8)$$

where η is water surface profile above still water level, h is the still water depth, D equals to $\eta + h$, g is acceleration of gravity, x, y are horizontal coorinate, t is time, M, N are x, y components of the momentum flux, and n is Manning's friction factor. The governing equations can be computed based on, for example, the Leap-Frog Method.

The governing equations for the computation model in this section are the basic equations of mass conservation and momentum conservation integrated over the vertical direction. The numerical model basically consists of four components, which are the basic equation, initial/source conditions, off-shore boundary condition, and run-up boundary conditions.

Figure 12.3 shows the total view of tsunami and storm surge simulations including risk analysis. The earthquake displacement model gives the important initial wave condition for tsunami propagation.

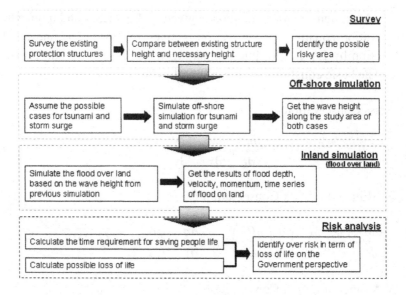

Figure 12.3 Tsunami and storm surge risk simulation.

12.3   Case Study, Indian Ocean Tsunami (2004)

Field surveys were performed on the disaster caused by the Indian Ocean
tsunami that occurred in December 26, 2004. Surveys were done in the
south part of Sri Lanka and in Aceh in Indonesia. From the survey
results, a variety of disaster mechanisms were found. Local natural
conditions and social conditions had a big influence on the disaster
mechanism. In order to promote an appropriate post tsunami rehabilitation
and environmental restoration process, cooperative work with local
engineers and local university professors is essential since the restoration
process should take into account the local social and natural conditions.

12.3.1   *Surveys in Sri Lanka*

Since the team of Prof. Fumihiko Imamura of Tohoku University
surveyed from Colombo to Gelle and the team of Prof. Philip L-F Liu
of Cornell University would survey the east coast of Sri Lanka, our YNU
(Yokohama National University) team (Shibayama *et al.* 2006) surveyed

the south coast of Sri Lanka. Figure 12.3 shows the summary results of measured tsunami run-up on the land along coast. Eight survey sites were investigated and the maximum run-up height in this southern area of Sri Lanka was found to be 10.6 m in Hambantota. Figure 12.4 shows the cross-sectional view of tsunami height distribution in the downtown area (east part) of Hambantota. In the area, the tsunami run-up height was not so high at 6.8 m. This means that the kinetic energy was not fully transferred to potential energy and the tsunami came to the downtown area with high velocity. This resulted that water with high momentum came to the area and produced very severe damage to houses and people in the area. Figure 12.5 shows the view of the west part of Hambantota. In the area, since there were coastal hills, the tsunami climbed the hill and its energy was transferred into potential energy. The maximum run-up height was 10.6 m in a hill on the west part.

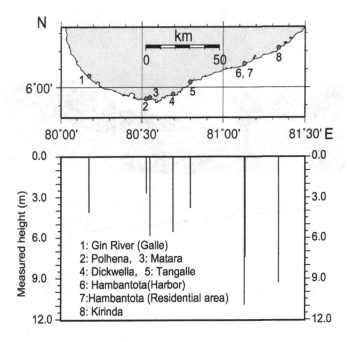

Figure 12.4 Summary of run-up in Sri Lanka.

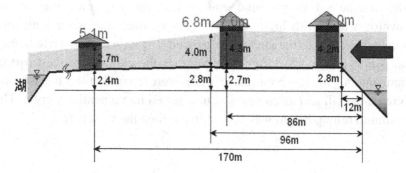

Figure 12.5 Tsunami height, east Hambantota.

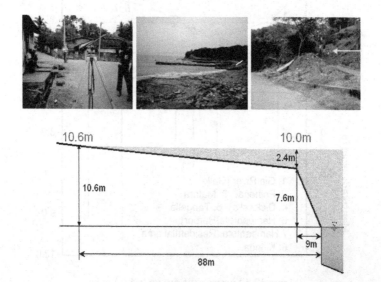

Figure 12.6 Tsunami height, west Hambantota.

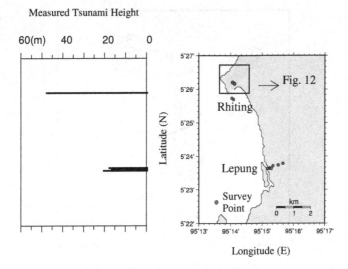

Figure 12.7 Tsunami run-up in Aceh.

12.3.2 *Surveys in Indonesia*

We also surveyed the Aceh area in Indonesia. Figure 12.6 shows the summary of the run-up height distribution along the west coast, that is, south from Aceh city. In Lepung, the tsunami height was more than 20 m (21.4 m). In this area, tsunamis still had high momentum and washed away the structures and trees on the surface of the ground. The measured height of tsunami was recorded in the coconuts trees, which survived the tsunami attack.

In Rhiting, the tsunami propagated over a hill in a peninsula. Figure 12.8 shows the topography of the peninsula. Tsunami energy was concentrated in a relatively low place of the hill and the tsunami flowed over the hill. Maximum run-up over the hill was 48.9 m and this is the maximum record of tsunami run-up in the Indian Ocean tsunami of 2004 (Shibayama *et al.*).

12.3.3 *Tsunami disaster prevention*

From the field survey, it appeared that the number of lives lost depended on the geographical and social conditions of local area. In Hambantota,

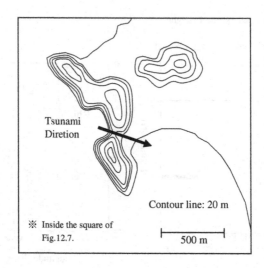

Figure 12.8 Topography in Rhiting.

a concentrated residential area was situated in a low elevated area between the sea and a lake. The tsunami flooded over the area with high momentum and swept out residents. In Lepung, the tsunami flowed over a fishery village and was reflected by the vertical wall formed by an adjacent hill and flooded again over the village. For the retreat and recovery process, we should consider the local condition of the topography and the social conditions of the area.

In order to establish a reliable disaster prevention system, we should design appropriate protection structures and also should design evacuation plan for residents in the area.

References

Shibayama, T., Okayasu, A. and Toki, M. (1990): Numerical simulation of storm surge in Tokyo Bay by using two level model, *Proc. of Civil Eng. in the Ocean*, JSCE, pp. 77-82 (in Japanese).

Tomoya Shibayama, Akio Okayasu, Jun Sasaki, Nimal Wijayaratna, Takayuki Suzuki, Ravindra Jayaratne, Masimin, Zouhrawaty Ariff and Ryo Matsumaru (2006): Disaster survey of Indian ocean tsunami in south coast of SriLanka and Ache, Indonesia, *Proc. of 30th Coastal Eng. Conf.*, pp. 1469-1476.

Subject Index

Author Index